WASTELAND MANAGEMENT AND ENVIRONMENT

Editors
A.K. Roy
S.K. Varma

SCIENTIFIC PUBLISHERS

Published by:
Scientific Publishers
5A, New Pali Road, P.O. Box 91
Jodhpur 342 001, India
E-mail: info@scientificpub.com
Website: www.scientificpub.com

Print : 2019

Disclaimer: Whereas every effort has been made to avoid errors and omissions, this publication is being sold on the understanding that neither the editors (or authors of chapters in edited volumes) nor the publishers nor the printers would be liable in any manner to any person either for an error or for an omission in this publication, or for any action to be taken on the basis of this work. Any inadvertent discrepancy noted may be brought to the attention of the publishers, for rectifying it in future editions, if published.

ISBN: 978-81-72331-47-4 (Hardbound)
ISBN: 978-93-87893-16-0 (E-book)

Printed in India

PREFACE

This book embodies the proceedings of the National Symposium on "Management of Wasteland to Protect Environment-Retrospect and Prospect" organised by the University Department of Botany on April 24-25, 1998. It was jointly sponsored by the Ministry of Environment and Forests and T.M. Bhagalpur University.

The focal theme of the symposium was to review the reasons for the generation of wastelands, measures for its prevention and reclamation, and impacts on the degradation of environment. The changing scenario leading to generation of wasteland due to deforestation, mining operations, development projects like construction of roads, dams, reservoirs etc., salinity alkalinity and water logging was discussed. Application of microbial technology to reclaim the degraded land was also reviewed.

The symposium was inagurated by Prof. S.R. Sirkar, Honourable Vice-Chancellor and presided over by Prof. S.K. Varma, Head Deptt. of Botany of the University. More than three dozen subject experts from different parts of the country participated in the academic deliberations. They reviewed various problems and prospects of wasteland management in relation to the protection of gradually degrading environment. They also expounded that the success of this programme requires a holositic approach and participation of ecologists, environment managers, sociologists, technocrafts, environment organisors and NGO's.

The proceedings includes both review and research articles on the various facets of the focal theme. Some articles are related to the reclamation of wasteland generated by mining operation while others are concerned with the prospects of microbial technology in reclamation of degraded lands. Articles on impact of industrialization and anthropogenic activities on aquatic and terrestrial environment are also included.

Broadly speaking this volume based on the proceedings of the symposium in its totally is expected to be of immense value to the scientists and researchers, working in the allied fields and also for the teachers and Post-Graduate students of Environment Biology, Taxonomy, Microbiology and Applied Botany.

The convenor of the symposium expresses his deep gratitude to the Vice-Chancellor Prof. S.R. Sirkar and Sri U.R. Biswas Conservator of Forests, for gracing the occasion. Besides, the convenor is also grateful to Prof. Anupam Varma, F.N.A., Dean, IARI, Delhi, Prof. J.S. Datta Munshi, Emeritus Scientist, (CSIR), Bhagalpur, Prof. G.P. Agrwal, R.D. University, Jabalpur and Prof. A. Mahadevan, Madras University for their good wishes for the success of the symposium. The convenor is also thankful to the colleagues of the Department for their ungruding help and to the Ministry of Environment and Forests, New Delhi for providing financial support for holding the symposium.

The convenor feels pleasure to record special thanks to all the contributors of this volume for their generously in accepting invitation and contributing the articles. It is the quality of their contribution that made this volume of profound academic interest.

The convenor is grateful to the Chairman and Recorders of the various technical sessions and to all the senior participants especially to Prof. Anil Kumar Head, Dept. of Geography and Prof. J.V.V. Dogra, University Dept. of Botany, who delivered informative and relevant key note address on this occasion. The convenor is sincerely grateful to Prof. S.K. Varma, Head Dept. of Botany, who, provided all the required facilities and encouragement for organizing the symposium.

The co-operation of my research students and PG students in the organisation of the symposium is duly acknowledged and I am personally thankful to all of them.

A.K. ROY
Convenor
National Symposium

LIST OF CONTRIBUTORS

S.K. Varma

University Deptt. of Botany
TM Bhagalpur University, Bhagalpur

U.R. Biswas

Conservator of Forest, Bhagalpur

A.K. Roy

University Deptt. of Botany,
TM Bhagalpur University, Bhagalpur-7

Rashmi Kumari

University Deptt. of Botany,
TM Bhagalpur University, Bhagalpur-7

S. Ram Reddy

Microbiology section, Botany Department,
Kakatiya University, Warangal - 506009

S.M. Reddy

Microbiology section, Botany Department,
Kakatiya University, Warangal - 506009

R.C. Rajak

Department of Post-Graduate Studies & Research in
Biological Science R.D. University,
Jabalpur-482001 (M.P.)

Kamal Prasad

Department of Post-Graduate Studies & Research in
Biological Science R.D. University, Jabalpur-482001 (M.P.)

A.K. Jha

Department of Botany, J.P. University, Chapra - 841 301

J.S. Singh

Department of Botany, B.H.U., Varanasi-221 005

M.S. Rao

Microbiology Section, Department of Botany
Kakatiya University, Warangal - 506 009

P. Pavan Kumar

Microbiology Section, Department of Botany
Kakatiya University, Warangal - 506 009

S.R. Reddy

Microbiology Section, Department of Botany
Kakatiya University, Warangal - 506 009

R.S. Singh,

Central Mining Research Institute,
Barwa Road, Dhanbad - 826 001

S.K. Chaulya

Central Mining Research Institute
Barwa Road, Dhanbad - 826 001

S.K. Singh

Central Mining Research Institute
Barwa Road, Dhanbad - 826 001

A. K. Kandya

Deptt. of Botany, Dr. H.S. Gour University, Sagar (M.P.)

Malaya K. Misra

Ecology and Floristics Laboratory
P.G. Department of Botany, Berhampur University
Berhampur - 760 007, Orissa

Kumar Birendra

Scientist, CMRI, Dhanbad-826001

R.S. Bilgrami

Deptt of Botany, BRA Bihar University, Muzaffarpur-842 001

S.K. Singh

University Department of Chemistry
T.M. Bhagalpur University, Bhagalpur

P.K. Mukherjee

Department of Botany, Calcutta University, Calcutta - 700 019

Arvinda Kumar Jha

Department of Botany
L.N. Mithila University, Darbhanga - 846 004

S.K. Goel

NEERI, Nagpur - 440 020

V. Jha

Deptt. of Biotechnology, LNMU - 846 004

S. Dutta

Department of Botany, Burdwan University
Burdwan - 713 104, West Bengal

P. Dutta Majumdar

Department of Botany, Burdwan University
Burdwan - 713 104, West Bengal

N.C. Chatterjee

Department of Botany, Burdwan University
Burdwan - 713 104, West Bengal

T.K. Pan

TM, Bhagalpur University, Bhagalpur-812 007 Bihar

R.P. Upadhyaya

TM, Bhagalpur University, Bhagalpur-812 007 Bihar

J. Duttamunshi

TM, Bhagalpur University, Bhagalpur-812 007 Bihar

S. C. Pal

P. I. Biomass Research Centre & Professor of Botany
Visva Bharti, Santiniketan - 731 235, W.B.

Anal K. Jha

University Department of Chemistry
T.M. Bhagalpur University, Bhagalpur - 812 007, Bihar

B.C. Mishra

Chief Technical Adviser, Golcha Group of Industries,
"Prem Prakash", S.M.S. Highway, Jaipur (Raj.).

Ambarish Mukherjee

Department of Botany, Burdwan University
Golapbag, Burdwan - 713 104, West Bengal

Debnath Palit

Department of Botany, Burdwan University
Golapbag, Burdwan - 713 104, West Bengal

K.R. Maurya

Bihar Agricultural College, Sabour, Bhagalpur, 813 210.

B.B. Mishra

Bihar Agricultural College, Sabour, Bhagalpur, 813 210.

R.B.P. Singh

En. Net Development Society, Dehradun

Piyoosh Rautela

En. Net Development Society, Dehradun

Dinesh Sati

En. Net Development Society, Dehradun

B.B. Mishra

Soil Survey and Land Use Planning Centre, Sabour, Bhagalpur, Bihar, 813 210

I.D. Singh

Soil Survey and Land Use Planning Centre, Sabour, Bhagalpur, Bihar, 813 210

J. Mall

Soil Survey and Land Use Planning Centre, Sabour, Bhagalpur, Bihar, 813 210

L.N. Singh

Soil Survey and Land Use Planning Centre, Sabour, Bhagalpur, Bihar, 813 210

S.P. Singh

Soil Survey and Land Use Planning Centre, Sabour, Bhagalpur, Bihar, 813 210

D.D. Pandey

Department of Botany, Nalanda College, Biharsharif - 803 101

Sanjeev Kumar

Department of Botany, Nalanda College, Biharsharif - 803 101

R. Prasad

Department of Botany, Banaras Hindu University, Varanasi-221005, India

B.K. Roy

Department of Botany, Banaras Hindu University, Varanasi-221005, India

P. Srinivas

Microbiology Section, Department of Botany Kakatiya University, Warangal - 506 009

G. Laxma Reddy

Microbiology Section, Department of Botany Kakatiya University, Warangal - 506 009

S.M. Reddy

Microbiology Section, Department of Botany
Kakatiya University, Warangal - 506 009

R.K. Gupta

Department of Botany, Koshi College, Khagaria

K.D. Pandey

Department of Botany, Banaras Hindu University, Varanasi - 221 005

A.K. Kashyap

Department of Botany, Banaras Hindu University, Varanasi - 221 005

Pankaj Kumar

NTPC, Kahalgoan,

Niraj Kumar

University Department of Botany
T.M. Bhagalpur University.

R.B. Prasad

S.M. College, Bhagalpur - 812 001

K. Sushma

S.M. College, Bhagalpur - 812 001

Kanak K. Pandey

Centre for Regional Study
TM Bhagalpur University, Bhagalpur-812 007

CONTENTS

CHAPTER - 1

WASTELAND MANAGEMENT - AN INTEGRATED APPROACH

S.K. Varma*

University Deptt. of Botany
TM Bhagalpur University, Bhagalpur

India has rich diversity of plants, animals and microbes which contribute significantly to the country's economy and environment. They not only provide food, medicine and raw material for industries but also help renewing the earth's atmosphere, absorbing pollution, enriching soil fertility and above all in maintaing the ethical and spiritual inspiration of the society. The rich heritage of plant life and the magnificent diversity of plant resources is being lost as forests and natural areas are cleared for urban, agricultural and industrial development or for meeting the needs of growing population. Each species and variety of plant represents a unique type of biological germplasm or gene pool with special characteristics and values. The preservation of plant genetic resource is, therefore, of prime importance and need to be addressed adequetly and on a priority basis. In this context Wastelands may be exploited for resource generation and for preservation and conservation of plant resources, as well as in protecting the environment and maintaing ecological equilibrium.

"As Chairman of the symposium it is my duty to explain the genesis for organising a National Symposium on Management of Wastelands to protect Environment".

---

* **CHAIRMAN** - "National Symposium "Wasteland Management"

Speaking of wasteland management now a days when almost every body is concerned and feels competent as wasteland manager is certainly a venture. Nevertheless, there is some justification for a naturalist to talk on the subject without repeating the views of ecologists, engineers, foresters, sociologist, politicians and all others who are already on the environmentalists arch.

"The beginning of all wisdom is to call things by their right names." Wastelands are regarded as ecologically unstable, with almost complete loss of top soil or have developed toxicity to life forms. These include areas affected by water logging, ravines, sheet and gully erosion, riverine lands, shifting cultivation, salinity and alkalinity, shifting sand dunes, wind erosion, extreme mositure stress and coastal sand dunes. The total wastelands in non forest area is estimated to be about 93.96 m ha in our country. Today a major global concern has been whether at the present rate of population growth there will be at the turn of the century enough land to meet the long term worldwide demands for food, fiber, fuel, fodder and shelter at reasonable cost. Experts are of the view that by 2000 AD India will require an additional 10 million hectares as net sown area, 40 million hectares for fuelwood and 10 million hectares for fodder. The increase of land resource is only possible through an integrated land use policy stressing both reclamation of degraded and under utilized land and prevention of land misuse.

The National wasteland Development Board has formulated programmes for the management and development of wasteland in the country through afforestation and tree planting. A tree is like a child and giving birth is only one part, often simplest and easiest. Nurturing the tree for the next 20 years while it matures takes the maximum effort. Secondly numerous examples across the country have shown the starting point of wasteland development is water and not trees. It is said that once a small water harvesting structure is built and an equitous system evolved for sharing the common water the people acquire a real interest in protecting and regreening the catchment of its water system. Water binds private fields, cattle and self interest to the health of the common land.

Therefore, planning for the wasteland development needs to be holistic and unified. Today planning is sectoral and fragmented. Wastelands authorities are interested in trees and not water, where as those interested in water are not interested in trees. Both, in any case, think big and not small so the result is pond without its protected catchment becomes nothing more than a hole in the ground, silted and dry. Trees keep getting planted, but they never survive.

The biggest failure of the last 50 years has been that the Government has tried to do everything itself from building factories and roads to planting trees and grasses. People have participated, but only when the Government has planned. What is clear is that India's vast wastelands cannot be managed by centralised bureaucracies.

Wasteland development demands democratic village level institutions that are in control of their own natural resource base. They must plant their own trees and benefit from them. Otherwise, planting trees on wastelands will remain an exercise in juggling numbers on paper.

From naturalist angle it is imperative to have some basic facts and information about different types of wastelands of our country, including their physical, biological and social health. Other information needed includes location, area and extent, rainfall pattern, soil health, geological makeup, vegetation and floristic composition and status of the wastelands private or Government? The second part concerns about management of the wasteland to meet the needs of local people, to restore ecological balance and improve environment, to improve hydrological regime of the area, to check soil and water erosion by adopting suitable techniques and to meet the needs of industries like paper and pulp.

In my opinion wastelands like abandoned mines, brick kiln, ravines and gullies, cultivated wastelands, fallows other than current ones, alkaline and other soils including coastal sandy areas may get priority for immediate reclamation.

Some of the priorities for integrated wasteland management are :

i. Establishment of a sound database

ii. Land suitability evaluation and soil capacity classification, taking into account soil characteristics and climate, topography, vegetation and socio-economic factors.

iii. Legistalive control of change in land use.

iv. Ameliorative and protective measures for wasteland management to minimise land damage.

Success of this programme depends on the involvement of the local people in addition to participation of plant scientists, ecologists, foresters, environment managers, sociologists, conservationists, engineers and technologists.

I hope that the views exchanged during these two days under four technical sessions will be fruitful, action oriented and aimed at providing maximum benefit to the common man. In doing so a balance between resource generation and utilization should be kept in view in-order to protect the environment through wasteland management.

CHAPTER - 2

ENVIRONMENT AND WASTELAND MANAGEMENT - PRESENT STATUS AND FUTURE PROSPECT

U.R. Biswas*

Conservator of Forest, Bhagalpur

Dictionary meaning : Wasteland means land not used for cultivation or building. In practice, Wasteland can be defined as "Areas where current biomass production seldom exceeds 20 per cent of its overall potential". Wasteland is degraded land. Of 320 million hectares of the country's geographical area approximately 158 million hectares are Wastelands.

Degradation of land takes place due to deforestation, unscientific cultivation, salinity, alkalinity, waterlogging and wind erosion etc. The degradation of environment in dryland areas is basically attributable to the increasing biotic pressure for the fragile ecosystems in the absence of adequate investments and appropriate management practices to augment and conserve the land and water resources.

Development activities like open cast mining, construction of a dam in power plants and slurry ponds in coal washery also degrade land. In the process of open cast mining, the overlying soil and rocks i.e. overburden is removed and stock-piled in dumps. The resultant overburden dump is thus, different from its original soil ecosystem, structure and horizonation. These dumps also lack in organic

* **PRESIDENT** - National Symposium on "Wasteland Management".

matter, useful plant nutrients and many microbial processes such as humification, soil aggregation, Nitrogen and Carbon cycling which are essential in the development and restoration of soil fertility and productivity. The nutritive and supportive capacity of this newly developed landscape is very poor and does not fully support vegetational growth. The dumps are also very prone to erosion problems due to high degree of slops and poor stability.

The country's current demand and supply gap in cases of timber, fuel wood, dry and green fodder is so large that it is difficult to bridge. The fodder shortage is also stupendous problems. Fuel wood, the primary sources of cooking energy for more than 70 per-cent of the population is obtained by gathering dead wood or from lopping of live trees. Leaf litters and twigs are also collected by villagers for cooking. This practice of collection of leaf litters result in deterioration of soil fertility due to lack of production of humus. Generally, the litter falling on ground form attractive food source to a variety of scavengers such as millipedes, woodlice etc. All these small creature pass the debris through their systems, the latter is broken into small pieces and the shredding process allows another group of fungi and bacteria to invade the litter. These organisms in turn continue to make the litter more palatable for a number of other decomposers. Release of nutrients from decaying organic matter in the soil is the critical step in ecosystem function. Fire is also a common cause of determination of land and its environment.

In 1984 the National Remote Sensing Agency reported that between 1975 and 1982 India has lost 1.3 million of forest annually.

The resulting environmental degradation is colossal.

1. Forest is gradually being depleted. Tropical forests are shrinking by 11 million hectares per year. In India it is estimated the annual denudation of forest land is between 1.3 million and 1.5 million hectares.
2. The top soil on crop land is rapidly eroding.
3. Deserts are expanding.
4. Lakes have dried up.
5. Under ground aquifer is not getting recharged.

In view of the problems discussed above, ways and means to improve and better manage the degraded land become extremely important. It is very necessary to in-creases nutrient level in the soil of degraded land. This can be attained through introduction of leguminous herbs and shrubs.

In order to overcome these problems and to rejuvinate the vast degraded lands, Forests Department of the state of Bihar has been implementing a number of schemes from its own plan funds. All these schemes focus on land and water resources management and afforestation.

Govt. of India also has from time to time taken up projects to develop and manage wasteland on watershed basis. The drought prone Area Programme, Desert Development Programme and Integrated Wasteland Development projects are three such projects. Some other programmes like JRY and Employment Assurance Scheme are also partly earmaked for overall management of wasteland.

While the Desert Development programme focussed on reforestation to arrest the growth of deserts, the DPAP concentrated on nonaerable lands and drainage lines for in-situ soil and moisture conservation, agro forestry, pasture-development, horticulture and alternative land uses. The IWOP made silvipasture, soil and moisture conservation on wastelands under Government, community or private control it's main purpose.

Other programmes aimed at improving wasteland in forest areas are Integrated Afforestation and Ecodevelopment projects and Area oriented fuelwood and fodder projects under NAEB. The long objectives of this projects are checking of degradation of forests and forest land, ecological restoration and environmental conservation. Short term objectives of these schemes are regeneration of degraded forests and adjoining area on water shed basis, agumentation of the availability of fuel wood, fodder and small timber, honey, wax fruits, nuts etc. and to secure people's participation in planning and regeneration efforts to ensure their sustainability etc.

We can not ignore human factor in any scheme under various programmes for management of wasteland and environment. Management of wasteland and its related environment involve the projects staff of implementing agency and the village community. The staff need training in technical matter as well as skill.

The villagers require training and exposure to modern scientific and technical methods, entrepreneurial skills to identify and explicit opportunities, community organisation and team building to work in groups.

Lastly, inorder to upgrade and improve upon quality of performance in the management of wasteland and environment suitable method of monitoring and evaluation has to be adopted to carry out concurrent as well as post-facto evaluations of the project. The result of these evaluations should be submitted to high level implementation and review committee with suggestion to policy issues and for improvement of management technique.

CHAPTER - 3

WASTELAND MANAGEMENT WITH SERICULTURE PLANTATION

A.K. Roy* & Rashmi Kumari
University Deptt. of Botany,
TM Bhagalpur University, Bhagalpur-7

INTRODUCTION

Decline in silk export graph in India and increasing demand of foreign currency it is required to look for the way towards the higher production of silk at lower cost. Further, the farmers in this region, generally do not raise their own plantation and depend solely upon nearby forest patches. Due to indiscriminate felling of tasar food plants for fuel purpose, they no longer have thick forest around them consequently they face many difficulties. So it is necessary to increase such forest with tasar food plants in systematic way to conduct rearing of tasar silk work. Our available land resources, those are not performing well due to embalances of biotic or abiotic factors so called degraded land must be exploited for sericulture plantation. However, the problem created by nutrient or water stress of soil may be tackled by the application of VA mycorrhizal fungi as biofertilizer during plantaton. In this direction, the suitable parameters have been proposed in the present article.

* **CONVENOR**- National Symposium on "Wasteland Management".

Silk Production and export potentials

Sericulture, the technique of silk production is one of the agro-based industries and it plays an important role in the improvement of rural economy. It has untapped export potentials and providing employment to nearly 6 million persons. Annual production of silk in the world is the estimated more than 45,000 tonnes of which Japan and China contribute more than 18,936 tonnes and 13,200 tonnes respectively. South Korea has also a significant place in sericulture. However, in India the production of raw silk of various texture i.e. Mulberry silk (2, 445 tonnes), Tasar silk (393 tonnes), Eri silk (91 tonnes), and Munga silk (41 tonnes) is to the tune of 3 thousand tonnes. the production data is very poor in comparision to other Asian countries.

Tasar plants growing regions

The forest of Madhya Pradesh, Orissa and Bihar have a good number of food plants of silkworm especially for tasar silk. These three states contribute major share of the total produciton of tasar silk. Madhya Pradesh alone contribute 45% of total tasar produciton in the country. Besides, maharastra, Andhra Pradesh, Uttar Pradesh and West Bengal are also engaged in cultivation of tasar silk and there is an immediate need for encouragement of such practices.

Varieties of silk/ silkworms and their food plants

Five varieties of silkworm are reared in India for producing the natural fibre silk, Bombyx mori, feeds on leaves of Morus sp. (*M. alba*, *M. indica*, *M. serrata*, *M. leavigata*) producing Mulberyy silk; *Antheria mylitta* feeds on *Terminalia* spp. (*T. tomentosa*, *T. arjuna*) and *Zizyphus mauritiana* producing Tasar silk; *Antheria royeli* & *A. peruyi* feeds on *Ricinus communis* producing Eri silk. Among these, the food plants of tasar silk worm have long term bearing of check green house emission effect, global warming, depletion of ozone layer, biodiversity and conservation.

Problems associated with cultivation of sericulture plants/ traditional practices of cultivation

In the rainfed plateau area of South Bihar rearing of tasar silkworm (*A. mylitta*) has already been taken on priority basis by the Tasar Research Board and local inhabitants using *T. tomentosa*,

(Asan); *T. arjuna* (Arjun) *Shorea robusta* (sal) and *Zizyphus mauritiana* (Indian plum) as host plants. However, the former two plants i.e. species of *Terminalia* are invariably cultivated in this area for the production of tasar silk cocoons. A usual method i.e. soaking of seeds for 48-96 hrs., sown separately in the poly-bags using desiered soil mixture (2 soil: 3 compost : 1 sand). After three months the seedlings are transferred to the field. NKP fertilizer is also being added when it is required. The Seedlings takes three years time to become suitable food plant for rearing silk worms. But most of these plants suffer from various kind of stresses like poor mobility of nutrients and water condition and their agronomic constrains, which together pose problem in establishment, growth and productivity of transplanted seedlings.

Constrains of wasteland development

Wastelands or degraded lands are synonymous with drastically distrubed lands where the native vegetation and animal communities have been removed and the top soil is lost, altered or buried. Such land can not be naturally rehabiliated within few years through normal successional process. In India, the problems of such lands is being increased day by day. About half of the total land area of India (175 mha. Approx.) has lost its usual productivity either due to natural or man made causes which poses a great threat to both the agriculture and environment simultaneously.

Exploitation of non-cultivable lands or wasteland resources needs a high priority. Advances in biotechnology have opened several new avenues for improving productivity of land adjusting different abiotic and biotic stresses. Current development in the field of applied soil microbiology and agro-forestry unveils great opportunities in optimum use of available resources and would need to increasingly emphasized in our efforts towards sustainability. Sericulture plantations especially tasar cultivation is one of the ideal tools for wasteland utilization and generation of plants wealth.

Exploitation of native VAM fungi in the cultivation of tasar plants on degraded lands

Many research findings have also confirmed that cocoon production can be increased significantly by doubling the leaf quality. It has been observed that by using improved technique 30,000 kg. Quality leaf can be produced per year at competitive costs against 15,000 kg. by adopting traditional methods in case of mulberry silk. Similarly tasar silk cocoons may be increasingly produced by using different innovative techniques of improving viability and capability of plants to produce more biomass. These plants have also successfully been raised in different polluted and wastelands of 'Korba' power city (M.P.) with various dose of NKP and FYM. The dose of costly chemical fertilizers may be lowered with the help of VA mycorrhizae. These VAM fungi have been proved from last twenty five years an efficient tool for improving the performance of plant even under different types of stresses especially water and nutrients. Dual inoculation of VAM fungi with other symbiotic (Rhizobium) and Asymbiotic (Azospirillum) bacteria have been proved more promising to improve plant performance.

The authors want to draw attention of researchers to the future prospects of these microbes in utilization for establishing the economically important plants on waste or degraded sites as these microbes may be helpful to -(i) enable the plant to survive even in the adverse climatic condition and nutrient stress of degraded lands consequently decrease rate of mortality; (ii) increase the growth rate and biomass of the plant so that the plant produces more quantity of leaves in lesser time and in turn supply more food (foliage) for insects and consequently in production of tasar silk fibre; (iii) to increase rearing potential on well maintained systematic plantation so that a rearer can easily get more cocoons.

So, the adoption of VAM technology will certainly have additional benefits like-(1) providing employment and promoting cottage based industry; (2) to set credit utilizing wastelands and of doing social welfare for supporting traditional tasar industry for the local inhabitants, which ultimately would change the economic condition of the rural people as well as the total scenario of the silk export of India.

Researches and development

During last twenty five years much attention has been paid to VAM fungi as a tool for improving the growth of agricultural and horticultural plants (Tinker, 1978; Gianinazzi *et al.*, 1989, 1990 a,b). The symbiotic system of these fungi has also been exploited to save the costly phosphatic fertilizers (Abbott and Robson, 1982 a,b). their importance in natural ecosystem (Koske, 1981) and in revegetation of the disturbed land has also been recognized (Reeves *et al.* 1979). Several reports are available which suggest that mycorrhizal inoculation stimulate rooting and growth and thereby transplant survival (Korenmanik *et al.*, 1978) of cuttings and seedlings raised in sterilized nursery media, is essential for the successful reclamation and revegetation programmes. It also helps in the establishment of plant cover on mine soil, eroded soil and industrial wastes, where the plant cover is difficult to establish (Hall and Armstrong, 1979; Hall, 1980).

The VAM infection has been reported to counteract with adverse soil factors. They alleviate heavy metals toxicity (Dueck *et al.*, 1986) and increase the tolerance of the crops to high acidity and temperature (Poss *et al.*, 1985).

The green house experiment conducted by Jasper *et al.* (1987) indicated that after 4-5 yrs. of revegetation programme the number of infective propagules appear to be restored to a level equivalent to that of undisturbed soil they have also given emphasis on investigating the possibilities to improve revegetation by increasing the inoculum potential of the disturbed sites.

Hayman suggested four major factors i.e. (i) the crop species involved (ii) the size and effectiveness of the indigenous VAM fungal population, (iii) the fertility of the soil (iv) and cultural practices can determine the success of inoculation. Pope *et al.* (1982) suggested the specificity of VAM fungi to different crop species. A number of International Organization viz. NIFTAL (Hawaii), ORSTOM (Senegal), USDA (USA), MRC (Kenya, Brazil, Bibans, Stockholm), ICARDA (Syria), IARI (Philippines), IIAT (Nigeria) are in use of biofertilizers. In addition to the International Programmes, various R & D Institutions and Universities are carrying out works on the plant growth stimulating aspects of mycorrhizae. It has been also shown that these fungi suppress many of the pathogenic microbes which cause failure of plantation on major scale (Rosendahl, 1985; Utkhede *et al.*, 1992; Baath *et al.*, 1984; Graham & Egel, 1988). However, only few information is available on application of VAM technology for the better cultivation of sericulture host plants. Positive effect of leaf quality on cocoon production was reported by Samasundaran *et al.* (1996). Mishra *et al.* (1996) demonstrated that add of VA mycorrhiza viz. *Glomus mossae* reduced the required phosphate level from 120 Kg to 30 Kg hac. in case of mulberry plant.

In India also several scientists are engaged in ecological physiological, biochemical, pathological, molecular and the allied areas of VA-mycorrhizal research work (Chandra & Kehri, 1990; Singh and Varma, 1987; Tilak & Negi, 1987; Vyas & Srivastava,

1987, Roy *et al.* 1997. Use of mixed inocula viz. dual inoculation of mycorrhizal fungi and N_2-fixing bacteria for afforestation programme is also well established. Singh and Subbarao (1987) and Singh (1987) reported increased growth of grasses when inoculated with VAM and *Azospirillum* sp. (Asymbiotic bacteria)

Different aspects of sericulture are being investigated in different research centres viz. CTRI Ranchi; PSPB, CSB (Varanasi); CSRI Berhampore; BSMTC, Chandrapur, RTRS, Bhandara, Silworm Seed Technology Lab., CSB (Bangalore); etc. Despite tremendous development in VAM technology to improve afforestation, fragmentary information is available on application of VAM as biofertilizer to rehabilitate wasteland for the purpose of sericulture. The vast streches of such lands are still deprived of such plantation. Wasteland Management Board, Govt. of India should have the priority for the reclamation of wasteland with tasar host plants which have dual benifits like increase in silk cocoon production and restoration of imbalanced environment.

REFERENCES

Abbott, L.K. and Robson, A.D. 1982a. The role of vesicular- arbuscular mycorrhizal fungi in agriculture and the selection of fungi for inoculation. Austr. J. Agric. Res. 33: 389-408.

Abbott, L.K. and Robson, A.D. 1982b. Infectrity of vesicular- arbuscular mycorrhizal fungi in agricultural soil. Austr. J. Agric. Res. 33: 1049-1059.

Baath E. *et al.* 1984. Trans. Mycol. Soc. 82(3) : 534-536.

Chandra, S. and Kehri, H.K. 1990. Microbial spectrum in mycorrhizasphere of green gram as affected by non-indigenous VAM fungi under normal and strained water schedules. In : Botanical Researches in India (Eds. Arya. N.C. and Chaudhary, B.L.), Himanshu Publications, Udaipur, pp. 514-521.

Dueck, T.A., Visser, P., Ennst, W.H.O. and Schat, H. 1986. Vesicular-arbuscular mycorrhizae decrease Zn toxicity to grasses growing in Zn polluted soil. Soil. Biol. and Bioch. 18: 331-333.

Gianinazzi, S., Glaninazzi-Pearson, V. and Trhuvelot, A. 1990. Potentialities and procedures for use of endomycorrhizas with special emphasis on high value crops. In: Biotechnology for fungi for improving Plant Growth (Eds., Whipps, J.M. and Lumsden, B.) pp. 41-54, Cambridge University Press, Cambridge, U.K.

Gianinazzi, S., Trouvelot A., Gianinazzi-Pearson, V. 1989. Conceptual approaches in agriculture for the rational use of VA-endomycorrhizae in agriculture: Possibilities and limitations-Agriculture, Ecosystems and Environments, 29: 153-161.

Graham J.H. & Egel D.S. 1988. Plant Disease 72(7): 611-614.

Hall, I.R. 1980. Growth of Lotus redunculatus Cav. in an croded soil containing soil pellets infested with endomycorrhizal fungi, N.Z.J. Agric. Res. 23: 103-105.

Hall, I.R. and Armstrong, P. 1979. Effect of vesicular- arbuscular mycorrhizae on growth of white clover, lotus and rye grass in some croded doil. N.Z. J. Agric. Res. 22: 558-608.

Jasper, D.A. Robson, A.D. and Abott, L.K. 1987. The effect of surface mining on the infectivity of vesicular-arbuscular mycorrhizal fungi. Austr. J. Bot. 35: 641-652.

Kormanik, P.P., Bryan, W.C. and Schultz, R.C. 1978. Endomycorrhizal inoculation during transplanting improves growth of vegetatively propagated yellow poplar. The Platn Propagator, 23 : 4-5.

Koske, R.E. 1981. A Preliminary study of interaction between species of vesicular arbuscular fungi iwa and dunes. Trans. Brit. Mycol. Soc., 76: 411-418.

Pope, P.E., Chney, W.R., Rhodes, J.D. and Woodhead, S.H. 1982. The mycorrhizal dependency of four hardwood three speacies. Unp. Bot. 61: 412-417.

Poss, J.A., Pond, E., Menge, J.A. and Jarrell, W.M. 1985. Effect of salinity on mycorrhizal onion and tomato in soil with and without additional phosphate. Plant & Soil. 88: 307-319.

Reeves, F.B., Wagner, D., Moorman, T. and Kiel, J. 1979. The role of endomycorrhizas in revegetation practics in the semi-arid west. I. A comparison of inicidence of mycorrhizae in severely disturbed natural environments. Amer. J. Bot., 66: 6-13.

Rosendahal S. 1985. Phytopathology 114(1): 31-40.

Roy, A.K., Rashmi Kumari, Singh, H.N.P. 1997. VAM association with kodo millet: A possiblility of its cultivation in rainfed areas. ISMN; 38, pp. 145-146.

Roy, A.K. and Rashmi Kumari 1999. In: Modern and innovatine approaches in soil management (Eds. D.J. Bagyaraj *et al.*) Rastogi Publication, Meerut, pp. 301-308.

Singh, C.S. 1987. Enhanced formation of vesicles, Arbuscules and spores through inoculation of grassess with Azospirillum brasileuse and Glomus Macrocarpum. Proc. Mycorh. Rround Table, New Delhi, 1987, 322-334.

Singh, C.S., and Subbarao, N.S. 1987. Yield and Phosphorus content of wheat (*Triticum aesticum*) as influenced by co- inoculation with Azospirillum brasilense and Glomus fasciculatum; Mycorrhi Round Table, Proc., 1987, New Delhi; 335-346.

Singh, K., & Varma, A. 1987. Mycorrhizal fungi, legume growth and root nodulatin in dry arid soil, Proc. Mycorrh. Round Table, New Delhi, 1987, 356-371.

Tinker, P.B. 1978. Effect of vesicular-arbuscular mycorrhizas on plant nutrition and plant growth. Physiol. Veg. 16: 743-749.

Utkhede R.S. *et al.* 1992. J. Phytopathology 135: 281-288.

Vyas, A., Shrivastava, H.P. 1987. Studies on vesicular arbuscular mycorrhizae of Moth Bean, Proc., Mycorrhiza, Round Talbe, New Delhi, 1987; 306-312.

CHAPTER - 4

REVEGETATION OF COAL MINE DISTURBED LANDS WITH MYCORRHIZAE- PROBLEMS AND PROSPECTS

S. Ram Reddy and S.M. Reddy

Microbiology section, Botany Department,
Kakatiya University, Warangal - 506009

INTRODUCTION

Mining activities generate a variety of wastes that differ in their biological, chemical and physical attributes. Factors such as soil fertility, pH, moisture, temperature and soil microbial activity influence successful plant establishment, growth and survival on such wastes (Visser, 1985). A myriad of tight or tenous links established between plants and soil micro-organisms, which have evolved *in situ* over long periods of time are broken. Coal mining spoils are poor habitats for plant establishment, growth and survival. They have low nutrient and organic matter content, low water holding capacity and low pH (Vogel, 1982). In addition, erosion and extremes in surface temperature are also impediments for revegetation. Mine spoils devoid of top soil have adverse chemical and physical properties that can persist for many years. Such sites are biologically unstable, and therefore, revegetation attempts are difficult. The primary objective of the mine waste reclamation include re-establishment of a sustainable plant community that is functional and beneficial to man directly or indirectly. This in turn demands the development of fully structural and functional soil microbial community (Take and Klein, 1985).

Functional soil microbial communities are composed of a complex of species that differ in their environmental tolerance, physical requirements and habitat adaptations (Perry *et al.*, 1987). The soil microbial community functions in a number of critical biogeochemical processes that are basic requirements for soil development and plant growth. A major component of the soil microbial community is vesicular arbuscular mycorrhizae (VAM) which form symbiotic association with roots of most land plants. They are ubiquitous both geographically and host-wise. The beneficial aspects of VAM fungi in plant growth under varied conditions has been elucidated (Sylvia and Williams, 1992). Despite the importance of VAM fungi in the revegetation of coal mine spoils, most current operational revegetation practices ignore VAM - plant community dynamics and have had limited success in establishing self sustaining plant communities. An urgent need, therefore, was felt to understand the VAM - plant community interactions on disturbed systems and to apply this knowledge in developing more innovative, cost effective and successful revegetation practices (Pfleger *et al.*, 1994). This review article addresses various aspects in applying the VAM technology in the revegetation of coal mine soil.

How Mycorrhizae Help in Sustainable Revegetation

Mycorrhizae become an important management issue in opencast mining because of three inherent factors associated with mining viz., removal of growing plants, soil disturbance, and topsoil storage can decrease the viable populations of mycorrhizal fungi (Jasper *et al.*, 1987). Revegetation efforts likely to remain incomplete and unsuccessful if the mycorrhizae are ignored. Revegetation of lands disturbed by mining is reported to have been facilitated by use of mycorrhizae (Lindermann *et al.*, (1984). Miller and Jastrow (1992) have identified the conditions in which management of mycorrhizae can become important in ecosystem restoration. Reeves *et al.* (1979) and Janos (1980) have suggested that the rate of succession or restoration may be hastened by inoculation or manipulation of the mycorrhizal fungus population, under conditions where VAM fungal propagule densities are very low. It has been observed that the mycorrhizal fungi prevent the stagnation of community development, but do not facilitate the skipping of successional stages. Many workers noted that severely disturbed lands under natural conditions are colonized by non-mycorrhizal or facultative mycorrhizal grasses (Miller, 1979; Allen, 1984 and Allen, 1984). A protracted occupation of a site by non-mycotrophic colonizers not only delay the establishment of more obligate mycotrophic plants but also result in a reduction in mycorrhizal fungal propagules reserves and hence, the continued occupation of a site by non-mycotrophic species (Allen and Allen, 1988 and Miller, 1987). Such spoils are usually characterised by having low species richness and often poor productivity. Hence, the primary goal of any restoration program should be to establish and maintain viable populations of mycorrhizal fungi. Studies of Fresquez and Linderman (1982) suggested that apart from VAM fungi introducing all components of soil microbial community back into the soil during revegetation is necessary.

Approaches in Revegetation of Coal Mine Spoils

Investigations of Johnson *et al.*, (1991) indicate that to accelerate revegetation of mine spoils it is necessary to use inputs that facilitate plant and microbial establishment and succession. Addition of amendments as an initial carbon and nutrient source is a means of quickly re-establishing the soil microflora and microfauna. Lindemann *et al.*, 1984) observed the addition of amendments to coal mine spoils will be necessary to establish and stimulate activity. They examined the effect of spoil amendments on soil microbial parameters and VAM fungal colonization of grasses. Their research revealed that the addition of either sewage sludge or hay increased the number of soil micro-organisms, enzymes activity and fungal genera distribution in non-rhizosphere spoil. Spoil plots amended with either mineral fertilizer, peat or liquid sewage sludge and seeded with slender wheat grass developed VAM formation after two weeks in peat-amended plots, 6-10 weeks after planting in fertilizer or sewage sludge amended plots. Colonization levels remained highest in the peat amended spoil after four growing seasons.

Since topsoils are the source of many beneficial soil micro-organisms including mycorrhizal fungi, replacement of top soil in the re-establishment of essential soil microbial community is considered as an alternate approach. Addition of top soil significantly hasten both the establishment of mycorrhizae and vegetation. In surface mining and in many landscaping operations, surface soils are commonly stripped and stockpiled in large mounds until they are needed for re-application during restoration. Many studies have demonstrated that this practice is detrimental to soil microbial community and physico chemical constituents (Harris *et al.*, 1989). Stockpiling reduces the mycorrhizal fungal densities. Rives *et al.*, (1980) found a significant reduction in the mycorrhizal infection potential (MIP) of soil stored for 3 years. This reduction in MIP levels associated with storage appears to be common to most mycorrhizal populations. In addition, stockpiling also reduces the effectiveness of biologically mediated processes such as soil aggregation and litter decomposition (Abdul-Kareem and Mc Rae, 1984 and Visser *et al.*, 1984). Nevertheless, top soil amendment is given priority in restoration of mine spoils. Lindermann *et al.*, (1984) demonstrated that top soiling at the depth of 30 cm increased the soil micro-organisms, enzymes activity and fungal genera. Of various amendments used, covering the spoil with top soil provides the greatest VAM colonization.

Field inoculations with VAM fungi is another way for restoration and revegetation of coal mine spoils. However, field inoculation with VAM fungi on a large scale is not currently practical. The use of nursery propagated VAM plants is one method of introducing suitable plant fungal combinations into mine wastes. This method will most likely be effective if the nursery plants propagated in the presence of ecologically adapted VAM fungi which are known to have a positive effect on plant survival and growth on mined sites. Roots of several woody plant species propagated in nurseries for outplanting in surface mines have been found to be colonized by VAM fungi. An example of this approach to revegetate has been demonstrated by Davies and Call (1987). They have observed not only the enhancement growth and development but also improved survival of the mycorrhizal transplanted seedlings. Non-inoculated plants of the same species transplanted on the same site showed very low levels of colonization even after two years.

There is also a good scope for inoculation of micropropagated tree seedlings with VAM fungi. This will be economical as much less inoculum is required for fortification of seedlings with symbiotically efficient fungi. The beneficial effects of VAM inoculation of micropropagated oil palm clones has been recently demonstrated by Blal and Gianinazzi-Pearson (1988). Experimental results by Revolanirina *et al.*, (1988) enlight the promising aspects of VAM tailored micropropagated explants of woody species in the restoration of vegetation on disturbed sites.

Because of depaupered VAM fungal densities in coal mine spoils, Allen (1984) and Allen and Mac Mahon (1985) suggested that a possible alternative for inoculating vast tracts of land would be to establish islands of mycorrhizal plants from which recolonization of the reclaimed spoil might occur more rapidly. Plants in the islands may also serve as catchments for VAM fungal spores that are wind blown onto a disturbed site as well as providing protective habitats for animals, insects that can carry spores and mycelial fragments.

Selection of VAM Fungi for Revegetation

Use of efficient VAM fungal strains is necessary for a spectrum of benefits to the plants. Criteria for the selection of efficient strains usually include : (i) ability of the isolate to survive and compete when inoculated; (ii) persistence of the isolate for a useful period in outplanted plant species; (iii) adaptibility of the isolate to a vide range of environmental conditions and (iv) a well developed capacity to form good hyphal system that augment the P uptake and also soil aggregate formation. VAM fungi are known to exhibit ecological specialisation. This fact highlights the importance of selecting VAM fungi which are ecologically adapted to a particular disturbed site for use in revegetation programs. Ecological specialization may be in response to several factors such as soil pH, presence of heavy metals, source of inoculum and host endophyte specificity. The literature is replete with information that different endophytes have different optimum ranges of pH and soil temperature, and also they differ in their adaptation or tolerance of phosphorus and nitrogen in soil/or plant (Mosse, 1990). Indigenous VAM fungi are likely candidates for reinoculation in reclamation effort. A number of workers (Lindsey *et al.*, 1977, Khan, 1981 and Stah *et al.*, 1988) employed indigenous VAM fungi as inoculants to study the effects on plant growth on mine spoils. Lambert and Cole (1980) suggested that the introduction of VAM fungi would be beneficial if indigenous fungi are absent or unadapted, available P is low, and plant used in revegetation positively respond to VAM.

Selection of Plant for Revegetation

Selection of plant type for revegetation play an important role in modifying the structural and functional aspects of the disturbed ecosystem. Miller and Jastrow (1992) have developed a conceptual model that depicts the role of plant roots and mycorrhizal fungal hyphae in promoting soil aggregation. Selection of C-4 plants which have high growth rate and release high levels of carbohydrates into the rhizosphere that stimulates microbial activity including VAM fungi is recommended for revegetation purpose (Rothwell and Eagelston, 1985). Lambert and Cole (1980) showed the importance of using the leguminous plants in revegetation. Leguminous plants, because of their ability to form tripartite symbiotic association, survive under minimal P level and faster rate of biomass production are considered as the potential candidates for revegetation for disturbed coal mine spoils. Facultative mycotrophs such as fourwing saltbrush (*Atriplex canescens*) are also ideal plants for revegetation purposes. This plant can establish and proliferates in the absence of VAM fungi, yet become mycorrhizal in the presence of VAM fungi. Ultimately, this plant species aids in the increase of the mycorrhizal inoculum potential levels (Aldon, 1978).

Epilogue

Though late, the importance of VAM fungi in revegetation efforts of various disturbed lands including coal mine soils has been realised and of late it has become an integral part of all afforestation programs. Although, studies involving the occurrence of VAM fungi on mining wastes are numerous, few studies have attempted to understand the interaction between VAM fungi and plant communities in revegetation success. There are many gaps in our understanding on practical aspects of application of VAM technology in revegetation of coal mine spoils. It is not clear what factors influence the rate of recolonization of coal mining wastes by VAM fungi. Some factors such as severity of disturbance, harshness of the site, low inoculum levels, edaphic characteristics, plant species, and time reclamation certainly influence the rate of VAM fungal colonization. There is a need for further experimentation to understand the influence of various factors on the colonization and reproductive potentials of VAM fungi. Of various prepositions of introducing VAM fungi, inoculation of seed, seed beds and nursery beds is a practical feasibility as it requires application of small quantities inoculum, for tailoring a large number of seedlings with mycorrhzae in a short span of time. The knowledge of VAM dependency of host species is crucial to a better understanding of plant nutrition and is considered necessary to predict the host response to VAM inoculation. Planting leguminous tree species in disturbed coal mine spoils is important for success of plantation as they are equipped with both rhizobia and VAM fungi. There are also reports that VAM fungi increase the population of associative nitrogen fixing bacteria of the genus *Azospirillum*. In view of the constraints in the production of mass inoculum efforts should be concentrated to develop suitable nursery technology for nodulation with rhizobia and fortification of seedlings with symbiotically efficient VAM fungi to produce quality seedlings for revegetation purpose.

REFERENCE

Abdul-Kareem, A.W. and S.G. Mc Rae, 1984. The effects on topsoil of long-term storage in stockpiles. Plant and Soil. 76 : 357-363.

Aldon, E.F. 1978. Endomycorrhizae enchance shrub growth and survival on mine spoils. In : The reclamation of disturbed arid lands, R.A. Wright ed., Univ. of New Mexico Press. Albuquerque. pp. 174-179.

Allen, E.B. 1984. VA mycorrhizae and colonizing annuals : Implications for growth. Competition and Succession. In : VA mycorrhizae and reclamation of arid and semi-arid lands (Ed. by M.F. Allen and S.W. Williams). University of Wyoming Agricultural Experiment Station Science Report SA 1261, Laramie, pp. 41-51.

Allen, M.F. 1984. Physiology of mycorrhizae : A key to understanding successful plant establishment. In : VA mycorrhizae and Reclamation of arid and semi-arid lands, S.E. Williams and M.F. Allen, eds., University of Wyoming Agricultural Experiment Station, SA 1261, Laramine pp. 69-80.

Allen, E.B., and M.F. Allen, 1988. Facilitation of succession by the non mycotrophic colonizer salsola kals (Chenopodiaceae) on a harse site : Effects of mycorrhizal fungi. American Journal of Botany. 75 : 257-266.

Allen, M.F., and J.A. Mac Mahon, 1985. Impact of disturbance of cold desert fungi comparative microscale dispersion patterns. Pedobiologia 28 : 215-224.

Blal, B. and V. Gianinazzi-Pearson, 1988. Second European Symposium on mycorrhizae. Prague, Czechoslovakia, p. 12 (Abstr.).

Davies, F.T. and C.A. Call, 1987. Survival and growth of mycorrhizal woody revegetation species in Texas lignite overburden. Page 148 In : Proc. 7th North American Conf. on mycorrhizae, D. Sylvia, L. Hung and J. Grahm, eds., Gainesville, FL.

Fresquez, P.R. and W.C. Lindemann 1982. Soil and rhizosphere micro-organisms in amended coal mine spoils. Soil Sci. Soc. Amer. J. 4 : 751-755.

Harris, J.A., Birch, P. and Short, K.C. 1989. Changes in the microbial community and physico-chemical characteristics of topsoils stockpiled during opencast mining. Soil use and management 5 : 161-167.

Janos, D.P. 1980. Mycorrhizae influence tropical succession. Biotropica 12 : 56-64.

Jasper, D.A., Rabson A.D. and L.K. Abbott, 1987. The effect of surface mining on the infectivity of vesicular arbuscular mycorrhizal fungi. Aust Jour. Bot., 35 : 641-652.

Johnson, N.C., Zac, D.R., Tilman, D., and Pfleger, F.L. 1991. Dynamics of vesicular arbuscular mycorrizae during old field succession. Oecologia 86 : 349-358.

Khan, A.G. 1981. Growth responses of endomycorrhizal onions in unsterilised coal waste. New Phytologist 87 : 363-370.

Lambert, D.H. and H. Cole, 1980. Effects of mycorrhizae establishment and performance of forage species in mine spoil. Agronomy Journal 72 : 257-260.

Lindermann, W.C., Lindsey, D.L. and P.R. Fresquez, 1984. Amendment of mine spoil to increase the number and activity of micro-organisms. Soil Sci. Soc. Am. J. 48 : 574-578.

Lindsey, D.L., Cress, W.A., and E.F. Aldon 1977. The effects of endomycorrhizae on growth of rabbitbrush fourwing saltbrush, and corn in coal mine spoil material. USDA Forest Service Research note RM-3431-6.

Miller, R.M. 1979. The ecology of vesicular arbuscular mycorrhizae in grass and shrublands In : Ecophysiology of VA mycorrhizal plants (Ed. by G.R. Safir), pp. 135-170. CRC press, Boca Ratan, FL.

Miller, R.M., and J.D. Jastrow, 1992. The application of VA mycorrhizae to ecosystem restoration and reclamation. Pages 438- 467 In : Mycorrhizal functioning and Integrative Plant Fungal process. M.F. Allen, ed., Chapman and Hall, New York, pp. 534.

Mosse, B. 1990. Mycorrhizal symbiosis and plant growth. Proc. of the second Nat. Conf. on mycorrhiza. In : D. Bagyaraj and Manjunath, A. (Eds.), Univ. Agriculture Sciences, Bangalore, pp 5-10 (Abstr.).

Perry, D.S., Molina, R., and Amaranthus, M.P. 1987. Mycorhizospheres and reforestation : current knowledge and research needs. Can J. For. Res., 17 : 929-940.

Pfleger, F.L., and R.G. Linderman, R.G. 1994. Mycorrhizae and plant Health. APS Press, The American phytopathological society, St. Pauls, Minnesote. pp. 1-344.

Ravolanirina, FS, Gianinazzi-Pearson, Trouvelot A. and M. Carre, 1988. Second European Symposium on Mycorrhizae. Prague, Czechoslovakia, p. 83 (Abstr.).

Reeves, F.B., Wagner, D., Moorman, T. and J. Kiel 1979. The role of endomycorrhizae in revegetation practices in semi arid west. I. A comparison of incidence of mycorrhizae in severely disturbed Vs natural environments.

Rives, C.S., Bajwa, M.I., Liberte, A.E., and R.M. Miller, 1980. Effects of topsoil storage during surface mining on the viability of VA mycorrhiza. Soil Science 129 : 253-257.

Rothwell, F.M., and D. Eagleston, 1985. Microbial relationships in surface mine revegetation. Environ. Geochem. Health 7 : 28-35.

Stahl, P.D., Williams, S.E., and M. Christensen., 1988. Efficacy of native vesicular arbuscular mycorrhizal fungi after severe soil disturbance. New Phytologist 110 : 347-354.

Sylvia, D.M., and S.E. Williams, 1992. Vesicular arbuscular mycorrhizae and sustainable agriculture. G.J. bethlenfalvay and R.G. Linderman, eds., ASA Special publication 54. Madison. WI. pp. 124.

Tate, R.L., and D.A. Klein, 1985. Soil Reclamation Processes, In : Microbiological Analyses and Applications, R.L. Tate and D.A. Klein, eds., Marcel Dekkar, Inc., New York and Basel. pp. 1- 349.

Visser, S., Fujikawa, J., Griffiths, C.L., and D. Parkinson, 1984. Effect of topsoil storage on microbial activity, primary production and decomposition potential. Plant and Soil. 82 : 41- 50.

Visser, S. 1985. Role of the soil invertebrates in determining the composition of soil microbial communities. In : Ecological Interactions in soil (Ed. by A.H. Fitter, D. Atkinson, D.J. Read and M.B. Usher). Special Publication of the British Ecological Society no. 4. Blackwell Scientific Publications, Oxford. pp. 297-317.

Vogel, W.G. 1982. A Guide for Revegetation coal mine soils in the Eastern United States. In : USDA for. Ser. Tech. Rep. NE-68, Northeast for. Exp. Stn. Broomhall, PA. pp. 1-190.

CHAPTER - 5

MICROBES AND WASTELAND MANAGEMENT: CHALLENGES AHEAD

R.C. Rajak & Kamal Prasad

Department of Post-Graduate Studies & Research in Biological Science R.D. University, Jabalpur-482001 (M.P.)

INTRODUCTION

Wasteland is an index of the country's use strategy of land resource. It is an indication of the level of ignorance and compulsions of people who are engaged in a perpetual war of attrition with land resource. This war finally ends up in the population of the village involved. Wasteland therfore, in Indian context can be divided into two broad categories: cultiviable and non-cultiviable. Cultiviable wasteland has potential for development for agricultural and pastoral purpose or can be afforested. It is not being used at present due to man constraints as lack of water, alkalinity of the soil, soil erosion, water logging or unfavourable physiographic position, or human neglect. Now, the problem is ever increasing due to excessive felling of forests, especially in ecological sensitive areas; overgrazing of the pastures; indiscriminate use of forested and fertile land for industiral and other development inappropriate agricultural practices all these lead to the creation and spread of wasteland.

Wasteland has recently attracted the attention of both government and public, because the land which is not suitable for agriculture can be utilized in a judicious manner for environmental protection by the process of land reclamation, soil conservation and afforestation. Although the present administration is having these devices for environmental modification and checking the spread of artidity and environmental degradation, but the progress has almost negative impact due to indifferent attitude of the government machinery to accomplish the task.

Wasteland Reclamation and Afforestation

Wastelands are synonymous with drastically disturbed lands where the native vegatation and animal communities have been removed and the top soil has been lost, altered or burried. Such lands will not become naturally rehabilitated within the life time of man through normal successional process. The rehabilitation process must start from a new root growth medium, from the artificial introduction of new plant communities, or by altering the conditions under which new plant communities may develop. Ideally, all the three efforts must be undertaken in order to expedite the rehabilitation. Examples of wastelands are surface mined lands for minerals (coal, kaoline, phosphate, bauxite, gravel, sand etc.) "borrow pits", saline soils, degraded material and wind and water eroded site.

To make these wastelands suitable for agriculture and forestry we have to pay attention towards physiography of the region, and conservation of the soil. Soil conservation can be achieved by mainly two methods :

(1) Small scale measures at local level and

(2) Major schemes of soil reclamation.

The soil of wastelands generally lack essential as well as accessary minerals and organic contents of these are in forms which is not easily available to the plant root as such. So the vegetative cover of such lands usually becomes very thin.

Micro-organisms may be very much helpful in the reclamation of wastelands soil. They play a significant role in producing green manure, biological control of disease and pest.

The different types of wastelands soil are generally deficient in phosphate, nitrogen and essential minerals, Mycorrhiza especially VA mycorrhiza, *Rhizobium*, *Azotobacter*, *Azospirillum*, *Nitrobacter*, *Azolla*, *Anabaena* etc. can play a significant role in achieving such important goals.

Benefits of Wasteland Management

Management of variosu type of wastelands through microbes helps in the beautification of environment, fodder, food, fuel and timber production; also in maintaining ecological balance hence the application of microbes (Biofertilizers) must be considered for realistic future development.

Mycorrhiza

Researches duing the first half of the present century have established that the mycorrhizal colonization is responsible to increase growth and nutrient uptake in the forest trees. These ecto and endo mycorrhizae (Table-1) are useful for reclamation of wasteland soils and to make them fertile (Draft and Nicolson, 1974; Daft *et al*., 1975; Daft and Haskaylo, 1976; Danielson *et al*., 1979; Jasper *et al*., 1987)

Table 1. Characteristics of important kinds of mycorrhizae in their natural state.

Character	VAM	Ecto	Ectendo	Arbutoid	Monotropoid	Ericoid	Orchid
Fungi	–	+	+	+	+	+	+
Septate	+	+	–	–	–	–	–
Aseptate	+	–	+	+	+	+	+
Hyphae enter cells	–	+	+or-	+	+	–	–
Fungal sheath	–	+	+	+	+	–	–
Hartingnet	–	+	+	+	+	–	–
Hyphal coil in cells	+	–	+	+	–	+	+
Haustoria dichotomous	+	–	–	–	–	–	–
Non dichotomous	–	–	–	–	+	–	+or-
Vesicles in cell or tissue	+or-	–	–	–	–	–	–
Achlorophylly	-or+	–	–	-or+	+	–	+

Importance of Ectomycorrhiza

- Enhances nutrient and water absorption through improve absorption surface area.
- Enhances nutrient mobilization through biological weathering.

- Increased in feeder root longevity by provision of a biological deterrent to root infection by soil born pathogens.
- Enhances uptake of elements such as N,P,K, Ca, Na, Fe, Zn, S and M_0 and their translocation to the host. Plants becomes more draught resistant.
- Many ectomycorrhizal fungi can degrade complex minerals and organic substances and make essential elements to the host plants.
- Mycorrhizal fungi afford protection to delicate root system from pathogenic fungi through the following means-

  a. Use of a surplus carbohydrates;

  b. Provision of a physical barrier;

  c. Favouring protective rhizosphere microorganism.
- Provide host plant with growth hormones including auxins, cytokinins, gibberellins, and growth regulating vitamins B.
- Important contributors to organic matter turn over and nutrient cycling in forest ecosystems.
- Ectomycorrhizal fungi members of Basidiomycotina, permeate deep to the forest floor and thus mobilize minerals in this zone, which subsequently are absorbed by the growings plants can be absorbed before they reach the sub-soil system.

Spore inoculum is prepared by blending freshly collected fruiting bodies of ectomycorrhizae (mushroom/puff balls) with tap water at high speed for two to three minutes. Some beneficial microoganisms within and on the surface of mature fruiting bodies of various ectomycorrhizal fungi have been found. Spores can be applied to the seed before sowing (Marx and Bell, 1985; Marx *et al*., 1984; Theodorum, 1984; Theodorum and Bowen, 1973) or 6-12 weeks after sowing either with a standard watering can or through the irrigation system. Spores of different fungi have been successfully used to stimulate growth of pines in Australia (Theodorum and Bowen, 1970, 1973; Theodorum, 1971) and South Africa (Donald, 1975). Marx (1976, 1980) and Ruchle (1980) have had similar success with inoculating *Pisolithus tinctorious* to assorted pines species in the United States. However, one of the major disadvantages of spore inoculum of most ectomycorrhizal fungi is the lack of appropriate laboratory test to determine spore viability. Another disadvantage is that sufficient sporophores of many fungi may not be available every year.

Another type of inoculm recommended for inoculation is pure mycelial or vegetative inoculum of ectomycorrhizal fungi (Trappe 1977; Marx, 1977 and Mikola, 1973). Molina and Palmer (1982) described isolation and maintenance of ectomycorrhizal pure cultures. Basically, a pure culture of particular fungus is obtained by isolating fungal material (Vegative tissue explant) in special media, to produce inoculum. Marx (1980) discussed in detail the early testing and development stages of producing viable inoculum of *Pisolithus tinctorious* for use in nursery and container grown seedling. Vermiculite based inoculum has been used successfully to form *P. tinctorious* ectomycorrhizae in fumigated soil on pine. Dramatic improvements in survival and growth of mycorrhizically tailored pine seedlings over naturally infected control seedling produced in containers or bare root nurseries have been reported from studies on acid coal spoils in Appacachia (Marx, 1977; Marx and Artman, 1977 and Walker *et al.*, 1980), Kaolin spoils in Georgia (Marx, 1977), severely eroded sites of the Copper Basin in Tennessee (Berry and Marx, 1978), borrow pits in South Carolina (Ruchle, 1980) and North Carolina (Goodwin, 1980) and prairie soil (Baer and Otta, 1981). Similar improvements in the performance of pine seedling have been reported on routine reforestration sites (Marx *et al.*, 1977; Mexal, 1980; Ruchle *et al.*, 1980 and Kais *et al.*, 1981).

VAM Fungi & Prospect in Reclamation of Wastelands

VAM fungi belong to Zygomycotina, order Endogonalses, family Endogonaceae. They are obligate symbionts and have not been cultured on any nutrient media (*In vitro* conditions) till dates and still discovery that fungal sporocarp and large distinctive fungal spores (100-600 m) found near or attached to mycorrhizal roots in the sterilized soil (Mosse, 1973; Gerdemann and Nicolson, 1963) and in axenic culture. A range of such resting spores different in morphological characteristics and life history now has been recovered from soils. The spores can be multiplied in presence of host plant and reproduce within 70-90 days, producing thousands of new spores of same kind.

Endogone resting spores are some of the large known fungal spores, which differ in shape and structure of the subtending hyphae, wall structure, cytoplasmic contents, colour, size and method of spore germination, morphology of secondary spores, presence or absence of sporocarps and details of spore arrangement within them.

There are only six genera of fungi are known to produce VA Mycorrhizae with plants. Two of the genera form clamydospores only. Four of the genera form spores that are similar to azygospores (*Gigaspora*, *Scutellospora*, *Acaulospora* and *Entrophospora*). VA mycorrhizae forming fungi are placed in the recently erected order Glomales or Achylosporaceae (Morton and Benny, 1990). There are several characteristics which the VAM forming genera sharing with each other (Table 2).

Table 2. General characteristics shared by genera with in Glomales of VAM

Genera	Soil Borne	Obligate	Sporocarp	Azygopore	Chlamydo spores	Septal perforation
Acaulospora	+	+	+	+	+	–
Entrophospora	+	+	–	+	–	–
Gigaspora	+	+	–	–	–	–
Glomus	+	+	–	+	+	+
Sclerocystis	+	+	–	+	+	+
Scutellospora	+	+	+	–	–	–

+ : Present; - : Absent

Uptake of phosphorus from soil

It has been demonstrated that mycorrhizal roots can absorb phosphorus from soil of low phosphorus status at a greater rate per unit length of root than non-mycorrhizal control. The external mycelium of the fungi is able to absorb 'P' away from the depletion zone around such root and translocate it to host cortex, but it is unlikely that mycorrhizal root have access to source of soil phosphates other than inorganic labile pool. It has also been mentioned by many workers that mycorrhizae make unavailable phosphorus available to the host plant (Swaminathan and Varma, 1979).

Other Roles of VA Mycorrhiza

The improved nutrient uptake of VAM is not confined to the phosphate alone. Infection can also help plants to overcome limitations imposed by the low mobility of other nutrient ions in the soil as Zn, Ca, Mg, Fe & S (Prasad and Rajak, 1998). At a particular soil phosphorus level mycorrhizal infection also increases nodulation and nitrogen fixation by legumes (Singh and Verma, 1980). VAM has shown an increase in stomatal conduction and photosynthesis during water stress in plant.

Microorganisms for Biological Nitorgen Fixation

Atmospheric nitrogen is fixed by some microorganisms enriching the soil and supplying this nutrient to the plants. The nitrogen fixation takes place through symbiosis between host plants and microorganisms or nitrogen is also fixed by free living microorganisms.

Rhizobium

Leguminous plants enter into a symbiotic association on their roots with N-fixing bacteria namely *Rhizobium* and *Bradyrhizobium*. Different species and strains of such bacteria are found to be effective in arranging a continuous supply of reduced nitorgen to the plants. These effective strains are cultured as mass culture under artificial conditions, and mixed in definite strength with carrier materials like charcoal powder and molasses etc. The formulation packed in sterilised packets are marketed for application in case of the specific crop either by treatment of the seeds of the legumes or sometimes by drenching of the base of the plant.

Mass Culture of Rhizobium

Many *Rhizobium* species grow and remain active in soil or rhizospheric nitrogen. Specific strains of *Rhizobium* are grown enmass in large vats containing sterilised Yeast mannitol agar broth and are suitably a mixed with charcoal powder or other carrier materials so that a minimum effective count is obtained. The powders are packed in small polythene backets of 200 g each for use as seed treating powders generally. Culture is often done in a fermenter desined suitably. Cultures for tree species used in social forestry, wasteland management etc., are also on their way to growers.

Azospirillum

Association of different species of *Azospirillum* with roots of some non-leguminous plants like Maize, Wheat, Sorghum etc. and nitrogen fixation by them has been received as a technology with great potentials.

Culture production

Selected cultures of *A. lipoferum* and *A. brasillense* is grown on potato infusion agar plates at 30 C for 72 hrs., scraped and suspended in dilute nitorgen free medium containing 1 g malic acid in 1 L of media. The suspensions so prepared are used for seed treatment. Six hrs dipping of seeds followed by 30 mins air drying are followed before sowing. Compared to crops without any fertilizer, treatment with *Azospirillum* increases grain yield 100%.

Cyanobacteria

Mass cultures of cyanobacteria for use as biofertilizer has been found to be possible in a mixed, open culture process. Large trays of galvanized iron of -8" depth and 6' x 3' dimention are placed outdoor. A thin layer of fresh garden soil is spread on the floor of the tray after mixing the soil with super phosphate. The bed is then carefully covered with a 3-4" layer of clean water without disturbing the soil layer. If required, suitable quantities of insecticide is also mixed with the soil. After 7-10 days the green thick scummy growth of the cyanobacteria in the bamboo baskets are used as manure in the wasteland management programes.

Azolla

Azolla is a floating fern commonly found in both warm and cold countries. Some species namely *A. caroliniana, A. micorphylla, A. pinnata, A. nilotica* etc. are common in different countries. They grow naturally in water-logged paddy fields.

The high rate of asexual and vegetative propagation of *Azolla* is utilized for its mass culture, on an average of 106 days *Azolla* contributes 120 Kg. nitrogen per ha. in addition to nearly 10 ton fresh weight of organic matter. Mass culture of *Azolla* is possible in paddy fields with standing water or shallow ponds and water bodies.

(i) For paddy fields 36 Kg *Azolla*, 300 g superphosphate and 36 Kg Furadan 3 G are introduced for every 100 sq. meter area. As soon as the field is covered with *Azolla* it may be ploughed under deeom posotion within 8-10 days.

(ii) In case of shallow ponds or masonry tanks with a mud-bottom, *Azolla* can be grown, harvested every 3-5 weeks and applied for the wasteland management. *Azolla pinnata* could be cultivated round the year i.e. having 22 crops in 335 days.

(iii) Growth of *Azolla* is favoured or influenced by several factors like atmosphere, air current, nutrient suply in water etc. *Azolla* grows most satisfactorily within a pH range of 5-9. For a better growth *Azolla* requires 20-30°C temperature of water although survived upto 14-40°C. Naturally in India more success in its cultivation has been achived in winter than summer. Higher light intensity (40-55 K Lux) inhibits growth of *Azolla*. The most desirable humidity for growth of Azolla is 85-90% and it dries up below 60%. A regular system of mass culture of *Azolla* in low lying paddy fields and in ponds, ditches and other water bodies under tree shade could be very useful in tropical areas.

Anabaena

The blue green algae *Anabaena* living in the dorsal surface of the leaves of *Azolla* and fixes atomospheric nitorgen. The *Anabaena* produces " Heterocysts" on growing older leaves of the *Azolla* the intensity increasing at the rate of nil on new leaf, 9.5% on 1st leaf and 25% on 2nd leaf where nitrogen fixation takes place. Actual blue green algae fixes 1-2 mole C_2H_2 / mg protein/ min when free and 5-7 mole as symbiont with the Pteridophyte. The nitorgen fixed as ammonia in leaf cavities is absorbed by the cavity hairs with the involvement of glutamic synthetase. Their symbiotic relation with BGA under Cyanophyceae contribute nitrogen to the paddy field. This is grown in rice fields in large scale in countries like China, Vietnam, Thiland etc. The BGA *Anabaena* grows in the cavities on the dorsal side of the leaves of *Azolla* and fixes atmospheric nitorgen which is aboserbed by the multicellular hairs of the cavity.

Importance of Microorganism

Due to limited application of nitrogen and phosphorus fertilizer in India only 15% of the potential yield of crops is achieved. Under the context micro-organisms have drawn greater attention in developing countries. Micro-organisms can bring down the requirement of import of fertilizers (656 tons of N and 142 tons of Phosphorus). Microorganisms (Biofertilizers) are biological in origin and serve to supply the nutrients for effective plant growth. Conservative estimates claim 20 million tons of nitrogen being fixed biologically every year by leguminous crops on about 250 million hac. area and this would be nearly 50% of the total nitorgenous fertilizer being produced in the world by fertilizer industries. The high demand for nitrogen as nutrient by the crop in general and high cost and difficult suply position for chemical nitorgen fertilizers, have created a situation in favour of biological nitorgen fixation technology.

CONCLUSION

The establ:shement of wasteland development in 1985 in New Delhi and wasteland corporation in West Bengal, 1987, the study of wasteland have gathered momentum in India. About 2.5 million hacteres of fertile land is being converted to degraded soils every year in India which inclues 1.5 million ha. of forest land. Plant of nitorgen fixing microbes and vesicular arbuscular mycorrhizae could help in achieving this goal. Since last four decades scientists have been trying to understand the varied and multilocational functions of mycorrhizae and other soil inhabiting microflora in soil-plant complex system. These studies have been simultaneously accompained by interrelationship studies, on mycorrhiza in association with soil mycoflora, asymbiotic and symbiotic nitrogen fixing micro-organism, including their role in rehabilitaiton of mine spoils, saline and alkaline soils. In addition, potential role of VAM has also been reported in eroded soil

This has suggested the reintroduction of VAM in eroded soils which help in the establishment of plants and also improve soil structure., thereby preventing soil erosion. Plant like *Alnus, Myrica, Casurina* and *Ceanothus* generally have a triparite associstion of the plant, nitrogen fixing bacteria and VAM fungi, which increases the rehabilitation of disturbed and degraded lands in moisture-deficient zones. Bloss (1988) emphasized role of VAM in improving the fertility of desert soil. According to Logan *et al.* (1989). VAM play an important role in sand dune management as they not only enhance the plant nutrients but also improve physical structure of soil. VAM and other microorganisms held tremendoùs potential for use in the agriculture and reclamation of wasteland where goals focus upon maximum plant establishment, growth and productivity at minimum cost.

REFERENCES

Baer, N.W. and Otta, J.D. 1981. Outplanting survival and growth of ponderosa pine seedlings inoculated with *P. tinctorious* in South Dakota. *For Sci.* 7: 277-280.

Berry, C.R. and Marx, D.H. 1978. *Pisolithus tincotrious* ectomycorrhiza on growth of lobolly and Virginia in the Tennessee Copper Basin. *USDA Forest Serv. Res. Note Se* - 264.

Bloss, H.E. 1988. Symbiotic microflora and their role in the ecology of desert plants. In: *Arid lands: Today & Tommorow* (Eds. *Whiteheed, E.E., Hutchinson, C.F., Timermann, B.N. and Varady, R.G.), Boulder, Colorado, USA, Westview Press Inc.* pp. 1031-1039.

Draft, M.J. Hacskaylo, E. 1976. Arbuscular mycorrhizas inthe arthracite and bituminous coal wastes of Pennsylvania. *J. Appl. Ecol.* 13: 523-531.

Draft, M.J. and Nicolson, T.H. 1974. Arbuscular mycorrhizas in plants colonizing coal wastes in Scotland. *New Phytol.* 73: 1129- 1138.

Draft, M.J. Hacskaylo, E. and Nicolson, T.H. 1975. Arbuscular mycorrhizae in plants colonizing coal spoils in Scotland and Pennsylvania. In : *Endomycorrhizal* (Eds. Sanders, F.E., Mosse, B. and Tinker, P.B.) Academic Press, New York and London, pp. 561- 580.

Daniclson, R.M. Zak, J. and Parkinson, D. 1979. Plant growth and mycorrhizal development in amended coal spoil material. In: *Ecology and Coal Resource Development* (Ed. Wali, M.K. Pergamon Press, New York, pp. 912-919.

Donald, D.G.M. 1975. Mycorrhizal inoculation for pines. *S. Afr. For. J.* 92: 27-29.

Frank, A.B. 1885. Ueberneue Mykorrhiza formen. *Ber. Bot. Gesell.* 5: 395.

Gerdemann, J.W. and Nicolson, T.H. 1963. Spore of mycorrhizal Endogone species extracted from soil by wet sieving and decanting. *Trans. Brit. Mycol. Soc.*. 46: 235-244.

Goodwin, O.C. 1980. Survival and growth of containerized Loblolly and Virginia pine seedling colonized with *P. tinctorious* on a borrow site in North Carolina. *W. For. Serv. Div. of For. Res. Releigh for Note.* 44. p.5.

Hirrel, M.G., Meharavaran, H. and Gerdemann, J.W. 1987. *Can. J. Bot. Soc.* 56: 2813-2817.

Jesper, D.A., Robso, A.D. and Abbott, L.K. 1987. The effect of surface mining on the infectivity of vesicular-arbuscular mycorrhizal fungi. *Austr. J. Bot.* 35: 641-652.

Kais, A.G., Snow, G.A. and Marx, D.H. 1981. The effects of benomyl and *P. tinctorious* ectomycorrhizae on survival and growth of long leaf pine seedlings. *Sough. J. Appel. For.* 5: 189-195.

Lakhanpal, T.N. 1988. *Proc. Mycorrhiza Symposium* Canada. (Eds. Verma, A. *et al), pp. 53-83.*

Lakhanpal, T.N. and Sharma, D.D. 1988. *Proc. Mycorrhiza Workshop, Canada (Eds. Verma, A. et al.)*, pp. 141-146.

Logan, V.S., Clarke, P.J. and Allaway, W.G. 1989. Mycorrhizae and root attributes of plants of coastal sand- dunes of New South Wales. *Austr. J. Plant Physiol* 61: 141-146.

Marx, D.H. 1976. Synthesis of ectomycorrhizae on loblolly pine seedling with basidiosporese of *Pisolithus tinctorious. For. Sci.* 22: 13-20.

Marx, D.H. 1977, The role of mycorrhizae in forest production. TAPPI Conf. Paper, Annual Meeting at G.A.

Marx, D.H. 1980. Role of mycorrhizae in forestation of surface mines. In: Proc. Symp. Tree for Reclamation of Interstate mining. Lexington, Kentucky.

Marx. D.H. and Artman, J.D. 1979. Pisolithus tinctorious ectomycorrhizae improve survival and growth of pine seedlings on acid coal spoils in Kentucky and Virginia. *Reclam. Rev.* 2: 23-31.

Marx, D.H. and Bell, W. 1985. Formulation of pisolithus ectomycorrhizae on loblolly pine seedlings with spore pellet inoculum applied at different times. *USDA Forest Serv. Res.* Paper 249.

Marx, D.H. Bryan, W.C. and Cordel,, C.E. 1977. Survival and growth of pine seedlings with Pisolithus ecotomycorrhizae after two years on reforesttration site in North Carolina and Florida. *For Sci.* 23: 367-373.

Marx, D.H. Cordell, C.E., Kenney, D.S. Mextal, J.G., Artman, J.D., Riffle, J.W. and Molina, R.J. 1984. Commercial vegetative inoculum of Pisolithus tinctorius and inoculation techniques for development of ectomycorrhizae on bare root tree seedlings. *For Sci.* 25: 1-101.

Mexal, J.G. 1980. Aspects of mycorrhizal inoculation to rforestration. *NZJ For Sci.* 10: 208-217.

Mikola, P. 1973. Application of mycorrhizal symbiosis in forrestry practices. In: *Ectomycorrhiozae : Their Ecology and Physiology.* (Eds., marks, G.C. and Kozlowski, T.T.) Academic Press, New York. pp. 383-411.

Molina, R. and Palmer, J.G. 1982. Isolation, maintenance and pure culture manipulation of ectomycorrhizal fungi. In: *Methods and Principles oj mycorrhizal research.* (Ed. *Schenek, N.C.). Am. Phytopathol. Soc.* St Paul, *Minn.* pp. 115-129.

Mosse, B. 1973. Advances in the study of vesicular arbuscular mycorrhiza *Anu. Rev. Phytopathol.* 11: 171-196.

Morton, J.B. 1988. Taxonomy of VA Mycorrhizal fungi: Classificaiton nomenclature and identification. *Mycotoxon* 32: 267-324.

Morton, J.B. and Benny, G.L. 1990. Revised classification of Arbuscular mycorrhizal fungi (zygomycetes). A new order Glomales, two new sub order Glominease and Gigasporineae and two new families Acaulosporaceae and Gigasporaceae with identification of Glomaceae. *Mycotaxon* 37: 471-494.

Prasad, K. and Rajak, R.C. 1998. Biotechnological application of mycorrhizae in reclamation of mined Dumps. *Proc. Integrated management of Plant Resources.* (In Press).

Ross, L. 1980, Growth of containerized lobolly pine with specific actomycorrhizae after 2 years on an anended borrow pit. *Reclamation Review* 3: 95-101.

Ruchle, J.L., Marx, D.H. and Abourough, M. 1980. Development of *P. tinctorious* and *Thelephoraterrestris* ectomycorrhizae on seedlings of coniferous trees important to Morocc. *Ann. de la Re cherche Forestriere.* Nu maroc. 21: 281-296.

Singh, K. and Varma A.K. 1980. Proc Mycorrhizal Workshop. Canada (Eds. Varma A. *et al.)* pp. 356-371.

Swaminathan, K. and Varma, B.C. 1979. Response of tree crop species to vesicular arbuscular mycorrhizal infection on Zinc- deficient Indian soil. New Phytol. 82: 481-487.

Theodorou, C. 1971. The phytase activity of the mycorrhizal fungus *Rhizopogen roseolous. Soil Biol. Biochem.* 3: 89-90.

Theodorou, C. 1984. Inositol phosphate in needles of *Pinus radiata* and the phytase activity of mycorrhizal fungi. *Being Proc. Congr. Soil Sci.* 3: 483-493.

Theodorou, C. and Bowen, G.D. 1970. Mycorrhizal responses of radiata pine in experiments with different fungi. *Aust. For.* 34: 183-191.

Theodorou, C. and Bowen. G.D. 1973. Inoculation of seeds and soil with basidiospores of mycorrhizal fungi. *Soil Biol. Biochem.* 5: 765-771.

Trappe, J.M. 1977. Selection of fungi for ectomycorrhizal inoculation in nurseries. *Ann. Rev. Phytopathol.* 15: 203-222.

Walker, R.F. West, D.C. and malanghlin, S.D. 1980. The dvelopment of ecotomycorrhizae on contrainerized sweet birch and European older seedling for planting on low quality soites. General Technical Report, *USDA Forest Service* 24: 409-417.

CHAPTER - 6

A LOW COST INPUT APPROACH FOR REVEGETATION OF COAL MINE SPOIL

A.K. Jha & J.S. Singh*

Department of Botany, J.P. University, Chapra - 841 301

*Department of Botany, B.H.U., Varanasi-221 005

INTRODUCTION

Degradation of ecosystem worldwide since 1945 has been estimated 17% of earth surface, of this, 750m. ha (38%) is slightly degraded (fully recoverable), 910 m.ha (46%) is moderately degraded (recoverable only through high inputs), 300 m. ha (15%) is severely degraded (recoverable with major interventions) and 9 m.ha (0.5%) is extremely degraded (non-recoverable). For India, estimates indicate that more than half of the land mass of the country is already degraded in some way, out of 143 million hectares of agricultural land (56%) is degraded due to bad agricultural practices.

The present stock of organic carbon in Indian soil (24.3 pg) can be increased to 34.9 pg by reforestation of multipurpose trees and grasses. It has been calculated that reforestation of 35 m ha of wastelands with suitable tree and grass species can sequester 0.84 and 1.06 pg of carbon in vegetation and soil, respectively (Gupta and Rao, 1994).

As per current estimates, 3-15 m.ha of tropical forests are cleared per year. Annual requirement of fuelwood, timber and paper pulp in India is 280 $m.m^3$, while the annual wood production in forest is only 52 $m.m^3$. This annual deficit of 228 $m.m^3$ is undoubtedly met through the consumption of huge capital. No wonder, in India the dense forest cover now is only 11% of the geographical area.

Features of the Northern coalfields limited (NCL) Singrauli

The Singrauli Coalfields, the future power capital of India, cover about 2200 km^2 in the Sidhi District of Madhya Pradesh and 80 km^2 in the Sonbhadra District of Uttar Pradesh. In about 300 km^2 area, about 10,850 million metric tons of coal reserves are present. NCL has acquired 19,352 hac. land, out of which 9508, 2716 and 6150 ha are forest, government and tenancy lands, respectively. There are elven collieries (Kakri, Bina, Marrak, and part of Khadia and Dudhichua in U.P. while Jhingwida, Gori, Moher, Amlori, Nigahi, Jayant and Part of Khadia and Dudhichua in M.P.) in NCL out of which nine are operational. About 90,000 metric tons of coal are excavated daily. Coal. reserves in different mines range from 14.67 to 348.93 million metric tons with striping ratio from 1:1.15 to 1 : 4.30, and the overburden ranges from 33.9 to 1,110 million m^3 (Singh *et al.*, 1995).

It is situated between latitude 23°41' - 24°12' N and longitude 81°48' - 82°52'E and elevation ranges from 280-519 m above mean sea level. Mean monthly minimum temperatures within the annual cycle range from 6.4-28°C and mean monthly maximum from 20-42°C. The annual rainfall averages 1069 mm. of which about 90% occurs during the period late June to September. The rainfall is characterised by a high degree of inter-annual variation.

Natural vegetation and wild life

Before coal mining and other industrial activities started in Singrauli area, the region was densely covered with tropical mixed dry deciduous forests. The common plant species were *Shorea robusta, Diospyros melanoxylon, Madhuca indica, Butea monosperma, Lagerstroemia parviflora, Holarrhena antidysenterica, Dendrocalamus strictus, Boswellia serrata, Wrightia tomentosa, Anogeissus latifolia, Woodfordia fruticosa, Emblica officinalis, Holoptelea integrifolia, Zizyphus mauritiana, Melia azedarach,* etc. (Jha and Singh, 1990 b).

Until 1950 the common wild life populations were of elephants, tigers and bisons. Sambhar, cheetal, neelgai and baikionj deer were in large numbers till 1965-66 when mining activity started.

Rapid industrialization and increased human activities recently have resulted in the converstion of substantial stretches of natural forests into savanna, grasslands, marginal croplands and barren lands. Surface mining for coal has caused extensive damage to the natural ecosystems with growing dumps of overbrden.

Mine spoils

Coal mine spoils are reclacitrant medium for plant and microbial growth because of low organic matter contents, unfavourable pH, either coarse texture or compacte structure. Thus spoils are physically, chemically, nutritionally and biologically improvised. Salinity, alkalinity, sodicity, acidity, poor water holding capacity and toxicity are major problems in revegetation processes (Jha and Singh, 1993b). Bradshaw (1983) has described the major problems of mine spoils and their short and long-term treatment.

It is essential to understand the structure (i.e. composition and species diversity, distribution pattern and biomass and nutrient stocks) and fuctioning (i.e. productivity, nutrient cycling, succession, competition, and immobilization and release of contaminants) of the degraded ecosystems (Singh and Jha, 1993).

Thus a study was conducted to understand the mechanism of natural revegetation process, on an age series of mine spoil (5-, 10-, 12- , 16- and 20 yr old) at Jhingurda Project, NCL, Singrauli. The changes in spoil physicochemical properties; microbial biomass C,N and P levels; species composition, community characteristics and plant biomass with age of mine spoils were assessed. The impact of age and different microsites (slope, coal patch margins, undulating surface and flat surface) were evaluated. Mine spoils were evaluated as a medium for plant growth, and potentially suitable plant species for revegetation purpose were screened

Physico-chemical properties

The proportion of coarse fragments (> 2.0 mm) was higher in mine spoils compared to forest soil. The proportion of coarse fragemnts decreasd with age of mine spoils ($r= -0.943$, $P <0.05$) and the proportion of 0.2-0.1 mm particles also increased with age ($r= 0.957$, $P<0.05$), (Jha and Singh, 1991).

Total soil N, mineral N, $NaHCO_3$ N, $NaHCO_3^-$ extractable pi and exchangeable k increased with the age of mine spoils (r= 0.930-

0.996, $P < 0.5$-0.01). These parameters in mine spoils were lower than in native forest soil even after 20 years of succession (Table 2). Exchangeable Na decreased with age of mine spoils ($r = -0.989$, $P<0.01$) and was higher than native forest spoil even after 20 years of succession.

The recovery of total soil N, a critical nutrient, during natural succession of mine spoils of 20 years of age was estimated to be 0.0012 per cent per year and it may take about 200 years of natural succession for total soil N pool to reach the native forest soil level (Srivastava *et al.*, 1989).

Soil microbial biomass C, N and P

Microorganisms play an important role in soil formation and revegetation through their activities as decomposers and nutrient cyclers, N_2 - fixers and mycorrhizal symbionts. Soil microbial biomass (a living part of the soil organic matter) is an agent of transformation of native organic matter and a reservoir for N, P and S (Jenkinson and Ldd, 1981). Plants serve as a carbon source for the microbial community and in turn microbes provide nutrients for plant growth through mineralization. Recognition of the importance of the soil microorganisms in the functioning of the ecosystems has led to an increased interest in measuring the nutrients held in their biomass (Powlson *et al.*, 1987; Singh *et al.*, 1989; Srivastava *et al.*, 1989).

The levels of microbial biomass C, N and P of an age-series (5-, 10-, 12-, 16- and 20- yr old) of coal mine spoils in a dry tropical environment at Jhingurda Project, NCL, Singrauli were measured (Srivastava *et al.*, 1989). Microbial C,N and P is mine spoil ranged from 209 to 496 $\mu g\ Cg^{-1}$, 20-36 $\mu g\ Ng^{-1}$, and 7-16 $\mu g\ Pg^{-1}$ soil, respectively. These values were maximum in forest soil and minimum in 5-yr old spoil. The biomass C,N and P were about four times greater in native forest soil than those in the 5-yr old spoil (Table 3).

MB-C ($\mu g\ g^{-1}$) can be predicted with some confidence ($r^2=0.993$, $P<0.01$) from the more easily measurable MB-P ($\mu g\ g^{-1}$) using the following equation :

$$MB - C = -17.49 + 30.47\ (\pm 1.06)\ MB - P$$

The soil microbial biomass can be characterized by a mean C:N:P ratio of 29 : 3 : 1 and is determined largely by total soil N(TSN) and $NaHCO_3$ pi ($R^2 = 0.976$, $P<0.05$) according to the following equation :

$$MB - C = -167.10 + 69.83\ pi - 104.35\ TSN$$

Where MB – C and pi are in $\mu g\ g^{-1}$ and TSN is in %.

Srivastava *et al.* (1989) hypothesized that soil microbial biomass is a critical factor in recovery of mine spoils as it aids in the reestablishment of nutrient cycling. The levels of biomass C, N and P can be taken as functional indexes of soil redevelopment.

Vegetation development

On four microsites, slope, coalpatch margin, undulating surface and flat surface on the basis of IVI values, nine communities were identified (Jha & Singh, 1992). *Dactyloctenium aegyptium* and *Digitaria setigera* were early colonizing community occurring on the 5-yr old slope microsite. The *D. aegyptium- A. adscensionis* community occurred on 5-yr old flat surface microsites. The late successional plant community *A-adscensionis* and *B. pertusa* first arrived on the 10-yr old undulating surfaces and continued to occur on this microsite up to the last spoil age (Jha and Singh, 1992). Plant species of later successional stage may be found from the beginning on mine spoils, but they grow rapidly after some measure of soil development have occurred (Bradshaw, 1983). *B. pertusa*, a species of higher successional order growing under better soil conditions, participated in community formation on mind spoils with *A. adscensionis*, which is a major species of degraded grasslands. This pattern of succession indicates the possibility of revegetation of the mine spoils with desirable species of higher successional order (Jha and Singh, 1992). The importance of microsites was evident in differences in spoil physico-chemical and biological properties, plant communities characteristics and plant biomass even after 20 yers of succession. Discriminant analysis indicated that undulating and flat surface microsites were close together. The slope and coalpatch margin microsites were very distinct from other microsites. Thus it was suggested that in any revegetation plan microsites should be taken into consideration, and the spoil dumps should be so structured as to have greater proportions of flat surface and undulating surface microsites. Natural colonizers should be taken into consideration in revegetation programmes, as they could hasten the natural succession process.

Assisted Vegetation and Soil Development

Natural vegetaion development is slow process because the

limiting factor is immigration of propagules, seeds, etc. Commonly it takes about 50 or 100 years for a satisfactory vegetation cover development on mine spoils (Bradshaw, 1997). However, in some situations natural colonization is a powerful tool. Normally selected species are planted or sown by using different techniques which ensures appropriate seed or plant placement. A desired range of species are introduced and problems which affect establishment of seedlings are overcome. In restoration of mine spoils, it is an important aspect of ecological engineering (Bradshaw, 1997). Human assistance is needed for rapid recovery of mine spoils.

Therefore, in about 14 ha area on mine spoils of Jayant and Bina Projects, a long-term minitoring experimental plots, having mosaics of vegetation covers consisting of ground vegetaion and tree species, slow growing and fast growing, were set up.

The major objectives of the study were (i) To know the physico-chemical properties of coal mine spoils of different collieries of NCL, Singrauli, and (ii) To develop ecotechnology for accelerating the revegetation of mine spoils by using biological resources. Long-term minitoring field experiments were setup at Jayant and Bina Projects in 1993. Following were the various revegetaion models:

(i) Tree monoculture seeded with grasses and legumes;

(ii) Tree monoculture with ground seeding and fertilizer applicaitons;

(iii) Tree monoculture seeded with crop plants;

(iv) Tree mixed culture seeded with grasses and legumes;

(v) Tree mixed culture with ground seeding and mulching.

In pot culture experiments suitable tree species for revegetation of mine spoils were screened.

The feasibility of direct seeding of tree species on flat and slope surfaces as a means of revegetation of mine spoils was evaluated.

For selection of suitability of plant species for revegetation of mine spoils, species characteristics such as adaptability of plant species to the particular physical and chemical conditions of mine spoils, and to macro-and micro-climatic conditions, drought tolerance, rooting depth, hardiness, palatability, availability of seeds, stabilization ability, asthetic appeal, ease of propagaton, and longivity of established plants were considered (Singh, *et al*., 1995). Choice of multipurpose trees can be of economic and social significance. Therefore, multipurpose trees should be given preference (Jha and Singh, 1994a). The plant architecture and geometry, morphology, phenology, root growth, architecture and activity are important parameters in selecting suitable tree species for revegetation purpose.

Pot culture experiments were conducted in polyethylene bags at Jayant and Bina projects nurseries during the period 1993-1995. Following treatmetns were given : (a) mine spoil only (control); (b) Mine spoil + 5 cm forest top soil; (c) Mine spoil + standard dose of NPK; (d) Mine spoil + 1/2 standard dose of NPK; and (e) Forest top soil only. Standard dose of NPK comprised: 120 kg N ha^{-1} as Urea; 60 kg ha^{-1} as single super phosphate; 60 kg N ha^{-1} as muriate of potash. Forty tree species were screened. Data for survival rate, shoot length, number of leaves and number of branches were collected periodically. Biomass values were quantified for root, stem and leaves.

Following thirtyfour species were found suitable for revegetation of mine spoils :

Acacia catechu (Katha, Khair),

Acacia nilotica (Babool),

Aegle marmelos (Bel),

Albizia lebbeck (Ksla siris),

Albizia procera (Safed siris),

Azadirachta indica (Neem),

Bauhinia racemosa (Kathmahuli),

Boswellia serrata (Salai),

Butea monosperma (Dhak),

Dalbergia sissoo (Sheesam),

Dendrocalamus strictus (Bans),

Diospyros melanoxylon (Tendu),

Emblica officinalis (Anwala),

Ficus benghalensis (Bergad),

Ficus glomerata (Gular),

Ficus religiosa (Peepal),

Gmelina arborea (Gamhar),

Grewia hirsuta (Gursakari),

Holoptelea intergrifolia (Chilbil),

Holarrhena antidysenterica (Koraya),

Lagerstroemia parviflora (Sidha),

Lannea coromandelica (Jigna),

Leucaena leucocephala (Subabul),

Madhuca indica (Mahua),

Melia azedirach (Bakain),

Morus alba (Shahtoot),

Pongamia pinnata (Karanj)

Shorea robusta (Sal),

Syzygium cuminii (Jamun),

Tamarindus indica (Imli),

Tectona grandis (Sagun),

Terminalia arjuna (Kahua),

Terminalia bellirica (Bahera),

Zizyphus jujuba (Ber)

In field trial experiments on 12-yr old mine spoil (flat surface) at Jhingurda, NCL, Singrauli (Jha and Singh, 1993 a) found the following 24 plant species suitable for revegetation of mine spoil:

Tree :

Pongamia pinnata, Acacia nilotica, Albizia amara, Albizia procera, Dichrostachys cinerea, Albizia lebbeck, Acacia auriculiformis, Acacia tortilis, Dichrostachys corten, Sesbania aegyptium.

Herbaceous legumes:

Phaseolus aureus, Cajanus cajan, Phaseolus mungo and *Pennisetum typhoides.*

Grasses:

Chrysopogon fulvus, Bothriochloa pertusa, Bothriochloa intermedia, Cenchrus setigerus, Dendrocalamus strictus

Spoil characteristics of different collieries of NCL

Spoil samples from 0-50 cm depth at 10 cm intervals (i.e., 0-10, 10-20, 20-30, 30-40 and 40-50 cm) were collected from spoils of different age (1 to 27 yr old) at Gorbi (10, 14 and 20 yr old), Nigahi (1 yr old), Amlori (2 yr old), Khadia (1 yr old), Dudhichua (2 and 5 yr old), Bina (1,4 and 12 yr old), Kakri (5 and 13 yr old), Jhingurda

(12, 23 and 27 yr old) and Jayant (1,4,6 and 8 yr old) collieries and from adjacent forest.

Among the collieries, peak values for pH total P, Cu, and Pb occured for the mine spoils at Jayant, for exchangeable Na at Bina, for exchangeable K, Ca, Mg, total Cd, Ni and Pb at Jhingurda. Amlori mine spoil showed maximum Cr. Gorbi maximum Fe and Zn and Kakri maximum Mn (Singh *et al*. 1996).

Spoil properties of various revegetation models at Jayant and Bina

The main values for pH, SAR (Sodium Absorption Ratio) and total Cd were similar at both the sites whereas Jayant spoil had greater values for exchangeable Ca, Na, total Cu, Cr, Pb, Ni, Fe, Zn and Mn and Bina spoil had higher values for exchangeable Mg, k and total P (Table 1).

Table 1. Properties of mine spoils of Jayant and Bina Collieries where long-term minitoring various revegetation models have been set up. The values are minimum and maximum obtained for 0-50 depth at 10 cm intervals.

		Mine spoil (0-50 cm depth at 10 cm intervals)					
		Jayant Project			Bina Project		
		Mini.	Maxi.	Mean	Mini.	Maxi.	Mean
pH	($\mu g\ g^{-1}$)	7.17	7.82	7.5	7.03	7.69	7.5
Ex Ca	(fgg g^{-1})	1672	3489	2448	1697	2311	2059
Ex Mg	($\mu g\ g^{-1}$)	6	20	12	14	18	16
Ex Na	($\mu g\ g^{-1}$)	575	1076	857	546	1089	776
Ex K	($\mu g\ g^{-1}$)	100	154	139	108	247	161
SAR	($\mu g\ g^{-1}$)	17	37	25	16	34	25
Total P	($\mu g\ g^{-1}$)	63	92	97	88	125	105
Total Cu	($\mu g\ g^{-1}$)	289	329	309	195	289	244
Total Cd	($\mu g\ g^{-1}$)	4	6	5	4	6	5
Total Cr	($\mu g\ g^{-1}$)	89	100	91	56	81	67
Total Pb	($\mu g\ g^{-1}$)	52	65	58	38	54	46
Total Ni	($\mu g\ g^{-1}$)	35	74	51	29	75	46
Total Fe	($\mu g\ g^{-1}$)	7832	21350	14013	6167	9838	8176
Total Zn	($\mu g\ g^{-1}$)	113	211	160	117	166	144
Total Mn	($\mu g\ g^{-1}$)	135	147	153	89	126	109

Tree Monocultre Seeded with Grassess and Legumes

Three monocultures were raised in mine spoils at Jayant and Bina Projects in about 2 ha area each in July, 1993. At Jayant Project nursery raised seedlings of following plant species were planted :

(a) Legumes :

Acacia catechu, Albizia lebbeck, Dalbergia sissoo, Leucaena leucocephala, Pongamia pinnata

(b) Non-legumes :

Azadirachta indica, Emlica officinalis, Gmelina arborea, Techtona grandis, Terminalia bellirica and *Dendrocalamus strictus*. Ground seedling was done with grasses *Pennisetum pedicellatum* and Heteropogon contorrus and Leguminous forb, *Stylosanthes hamata*.

At Bina Project nursery raised seedlings of following plant species were planted :

(a) Legumes :

A. lebbeck, D. sissoo, L. leucocephala, P. pinnata

(b) Non-legumes :

A. indica, D. strictus, Holoptelea integrifolia, Mddhuca indica, E. officinalis, Tamarindus indica, and *Terminalia arjuna*.

Stylosanthes humilis a leguminous herb was seeded in the plantantion plots.

In mine spoils of Jayant Project, growth performance data collected after 3 years of plantations are alike Singh *et al*. (1996).

At Jayant project, growth of leguminous trees in terms of height ranged from 2.49 to 3.24 m (x = 2.8m) diameter growth from 3.02 to 6.01 (x= 4.6 cm) and tree volume from 2317 to 11703 (x= 6526 cm^3). These values for non-legumes were : height 1.07 to 3.05 m (x= 2.2 m), diameter 2.52 to 6.28 cm (x = 4.1 cm) and tree volume 1919 to 12029 (x = 4276 cm^3). Leguminous trees showed better growth performance compared to non-leguminous species.

Similarly, at Bina Project also leguminous species showed better growth performance compared to non-leguminous species.

Tree monoculture Seeded with Crop Plants (agroforestry model)

In about 2 ha area in Jayant Project in July, 1993, *Pennisetum typhoides* was sown in the plots of leguminous trees (*Acacia catechu, Albizia lebbeck, Dalbergia sissoo, Leucaena leucocephala* and *Pongamia pinnata*) and *Cajanus cajan* was combined with non-leguminous tree (*A. indica, C. arborea, T. bellirica* and *T. grandis*).

The objective for setting up of this revegetation model was to rapidly increase the organic matter inputs and soil fertility and to check the soil erosion. *Cajanus cajan* and *Pennisetum typhoides* grow well on mine spoils (Jha and Singh, 1993 b). *Cajanus cajan* not only fixes nitrogen, but also increases total phosphorus availability due to the presence of VAM fungi in roots. Therefore, *Cajanus cajan* can grow and yield well in soils of low available P level and without applications of P fertilizer because of its ability to tap Fe - P, which cannot be easily utilized by the other crops (Ae *et al*., 1990). *Cajanus cajan* is a good soil binder also.

Above ground biomass of *C. cajan* averaged 1065 gm^{-2} and that of *P. typhoides* 672 gm^{-2}. The grain yield of *P. typhoides* was 176 g m^{-2} and that of *C. cajan* 92 g m^{-2} (Singh *et al*., 1995). Growth performance of leguminous trees; height, diameter and tree volume ranged from 2.84 to 5.31 (x = 3.9 m), k.70 to 9.50 (x = 7.3 cm) and 9227-39800 (x= 2354 cm^3), respectively. Whereas these values for non leguminous trees recorded ranged from 1.71 to 3.40 (x = 2.6 m), 4.80 to 8.50 (x = 6.1 cm) and 3940 to 24565 (x= 11214 cm^3), respectively, for height, diameter and tree volume. Leguminous species showed better growth performance than non-leguminous species.

Tree Mixed Culture Seeded with Grasses and Legumes

Mixed plantation were raised in about 2 ha each in Jayant and Bina Projects. Following were the tree combinations at Jayant Project :

Acacia catechu + Pongamia pinnata,

Albizia lebbeck + Acacia catechu,

Azadirachta indica + Emblica officinalis,

Dalbergia sissoo + Tectona grandis,

Dendrocalamus strictus + Tectona grandis,

Gmelina arborea + *Pongamia pinnata*,

Pongamia pinnata + *Terminalia bellirica*,

Terminalia bellirica + *Gmelina arborea*,

Tectona grandis + *Leucaena leucocephala*.

In between rows of plantations, seeds of grasses (*Pennisetum pedicellatum* and *Heteropogon contortus*) and a leguminous forb (*Stylosanthes hamata*) were sown.

At Bina project following tree combinations were raised and in between rows, seeds of a leguminous forb, (*Stylosanthes humilis*)

A. indica + *A. procera*,

A. procera + T.indica,

C. fistula + *M. indica*,

H. integrifolia + *D. sissoo*,

M. indica + *H. integrifolia*,

E. officinalis + *A. procera*,

P. pinnata + *D. strictus*,

T. bellirica + *A. procera*,

T. arjuna + *P. pinnata*, and

T. indica + *P. sissoo*

The objective for getting up of mixed culture plantation was to develop multistratal vegetation cover on mine spoils, which will check soil erosion effectively and will increase soil fertility status quickly.

At Jayant Project after about 3 yrs of plantations, greatest height was attained by *D. sissoo* in combination with *T. grandis* and the least height by *T. bellirica* in combination with *G. arborea*. Diameter was heighest for *G. arborea* in combination with *P. pinnata* and least for *T. bellirica* in combination with *G. arborea*. The number of culms/ clump in *D. strictus* in combination with *T. grandis* ranged from 2 to 7. The maximum height ranged from 1.75 to 3.75 m (x = 2.70 m) and maximum diameter growth varied from 2.86 to 1.68 cm (x = 2.4 cm).

At Bina project, greatest height was attained by *H. integrifolia* in combination with *D. sissoo*, x and the least height by *C. fistula* in combination with *M. indica*. Diameter was highest for *T. indica* in combination with *D. sissoo* and least for *C. fistula* in combination with *M. indica*

Effect of mulching on growth of trees grown as mixed culture with ground seeding

Grass hay was spread in between the rows of plantations in mixed culture plots, the driest site where tree growth was low, in December 1994 and data for growth performance of tree species were recorded in April, 1996.

Significant per cent increase in height and diameter growth after mulching in plant species have been observed. The increase in height growth was recorded higher in *Acacia catechu, Dalbergia sissoo* and *Terminalia bellirica* compared to *T. grandis, P. pinnata, A. indica, A. lebbeck, G. arborea, E. officinalis*, and *L. leucocephala*. In terms of diameter growth, greater effect of mulching was recorded for *A. lebbeck, D. sissoo, A. catechu, L. leucocephala, T. grandis, P. pinnata* and *A. indica* compared to *G. arborea T. bellirica* and *E. officinalis*.

Effect of NPK fertilization on biomass yield of *Pennisetum pedicellatum*

In Jayant project, plots of the followign tree species were seeded with grasses (*Pennisetum pedicellatum* and *Heteropogon contortus*) and the leguminous forb (*Stylosanthes hamata*): *Acacia catechu, Albizia procera, Azadirachta indica, Dalbergia sissoo, Dendrocalamus strictus, E. officinalis, Gmelina arborea, Pongamia pinnata, Tectona grandis* and *Terminalia bellirica*. Three replicate plots of 20 m x 20 m size for each species were maintained in this experiment. Each 20 m x 20 m plot was divided into four sub-plots of 10 m x 10 m size. One sub-plot was treated with full dose of NPK, another sub-plot with half dose of NPK and two sub-plots were maintained as control (without fertilizer). In each sub-plot 20 cm wide and 10 cm deept furrows were dug in between the planted seedlings. Furrows were made in vertical and horizontal lines. In vertical furrows seeds of grassess were sown and in horizontal furrows seeds of leguminous forb, *Stylosanthes hamata* were sown. Seeds of *Stylosanthes hamata* were scarified in hot water for few minutes before sowing. First of all the fertilizer mixture was spread in furrows and then sowing of grasses and *Stylosanthes hamata* seeds was done, and the furrows were covered with mine spoil. The seedling rate for grasses and leguminous forb was 7 kg ha^{-1}.

Full dose of fertilizer for grasses was : N as urea, 60 kg/ ha; P as single super phosphate, 30 kg/ ha; K as muriate of potash, 40 kg/ha. In the case of legumes the full dose was: N as urea, 20 kg/ ha;

P as single superphosphate, 40 kg/ ha; K as muriate of potash, 40 kg/ha.

Result indicated a marked effect of NPK fertilizer on dry matter yield of *P. pedicellatum* at both the sites. However, greater plant biomass occurred on the less rocky site than on the more rocky site Singh *et al.*, 1996. Shoot biomass was four times as large in 1/2 dose treatment but only five times as large in full dose treatment as in control. The root biomass on less rocky site in control was 19% and 25% ofthat of the full dose and 1/2 dose treatments, respectively. On the more rocky site the corresponding values were only 14 and 20%. Thus the relative response of root system to fertilizer was greater on more rocky site.

Growth performance of tree species in various revegetation model plots

When the data collected for growth performance of tree species in various revegetation models were compared, in general, it was found that the better growth performance in terms of height growth, diameter growth and tree volume, was observed in tree monoculture seeded with crops compared to tree monoculture and tree mixed culture with ground seedling with grasses and legumes.

In the experimental plots set up at Jayant and Bina projects growth performance data were recorded better for Jayant project compared to Bina project (Table-2). Jayant mine spoils are better growth medium for plants compared to Bina mine spoil. When direct seeding of tree species on flat and slope area, the data for growth performance are compared it is evident from the findings that the potentially suitable species are *A. catechu*, *L. leucocephala*, *Azadirachta indica*, etc.

Recuritment of plant species in revegetation plots at Bina and Jayant Projects

The revegetaion plots were set up in July 1993 at Bina (4 ha) and Jayant project (10 ha). The naturally colonizing plant species were collected after 3 yrs and were indentified (Table 3).

At Bina project ten tree and shrub species colonized the revegetation plots whereas the Jayant project seventeen tree and shrub species were recorded. The total number of grasses and herbs were 30 and 50 at Bina and Jayant projects, respectively (Table 3). The number of individuals of each tree and shrub species was higher at Bina compared to Jayant because at Bina intact forest area is located adjacent to the revegetation plots which helped in rapid

Table 2. Comparison of growth performance of plant species in various revegation models set up at Jayant and Bina Projects, NCL, Singrauli.

Revegetaion Models	Projects	Height growth (m)		Diameter growth (cm)		Height:diameter ratio		Tree volume (cm)	
		Maximum	Minimum	Maximum	Minimum	Maximum	Minimum	Maximum	Minimum
1. Tree mono culture seeded with grasses and legumes	Jayant	3.24 *A.lebbeck*	1.07 *T. bellirica*	6.28 *G.arborea*	2.52 *D.strictus*	110 *D.strictus*	25 *T. bellirica*	12029 *G. arborea*	1918 *D. strictus*
	Bina	2.07 *D.strictus*	0.43 *M.indica*	2.76 *T.arjuna*	0.48 *H.integri-folia*	177 *H.integri-folia*	45 *M.indica*	1249 *T. arjuna*	20 *H.integri-folia*
2. Tree mono culture seeded with crop plant	Jayant	4.31 *L.leucoce phala*	1.17 *T.bellirica*	9.5 A.lebbeck	4.8 *T.bellirica*	1.73 *L.leucoce-phala*	36 *T.bellirica*	39800 *A.lebeck*	3940 *T.bellirica*
3. Tree mixed culture seeded with grasses and legumes	Jayant	4.42 *D.sissoo*	0.92 *T.bellirica*	6.1 *G.arborea*	2.3 *T.bellirica*	111 *D.strictus*	4038 *T.bellirica*	9637 *G.arborea*	750 *E.officinalis*
	Bina	1.44 *H.integri-folia*	0.43 *C.fistula*	2.57 *T. indica*	0.70 *C. fistula*	74 *A. indica*	46 *A. procera*	2965 *R. procera*	21 *C. fistula*
4. Direct seeding of tree species on flat surface	Jayant	2.37 *L.leucoce-phala*	0.83 *T. arjuna*	4.13 *A. actechu*	1.62 *T. arjuna*	109 *Z. jujuba*	31 *S. cumini*	3650 *A. catechu*	218 *T. arjuna*
5. Direct seeding of tree species on slope	Jayant	2.61 *A.indica*	0.92 *S.cumini*	5.52 *A.catechu*	2.03 *D. leucoce-phala*	115 *L. leucoce phala*	30 *T.bellerica*	9602 *L. leucoce phala*	478 *T. arjuna*

recuitment of woody species. After 20 years of natural revegetation at Jhingurda project total number of plant species was 81 whereas only after 3 years of setting up of revegetation plots, the total number of species was 40 and 67 at Bina and Jayant projects, respectively.

Table 3. Total number of vascular plant species colonised in various revegetation model plots at Jayant and Bina projects and during natural revegetation at Jhingurda project, NCL, Singrauli (Based on Jha and Singh, 1990, 1991, 1992; Singh, *et al.*, 1996).

	Jayant Project (in 3 yrs after Plantation)	Bina Project (in 3 yrs after Plantation)	Jhingurda Project (in 20 yrs after natural succession)
Shrub & Trees	17	10	11
Grasses & Herbs	50	30	70
Total	67	40	81

After three years of raising of revegetaion models, a variety of wild life have inhabited the plots and it is not easy and safe in visiting the plots alone. Even forest department personnel hesitate in entering the model plots who are engaged in protection of plots (Singh, 1996).

Evaluation of direct seedling Technique

Direct seeding could prove to be a cost effective technique for revegetaion of mine spoils both in plain and slopy areas. Large area can be seeded in less time with less manpower, and seedling can be accomplished when and where planting of seedling is not feasible. Direct seedling often results in a more complete occupation of the site (Singh, *et al.*, 1996). Thus the trials were made to asess the relative success of this technique with available tree seeds sown on flat surfaced and sloping mine spoils.

In about 1.5 ha area on flat surfaced mine spoil at Jayant project ten tree species (each in 20 m x 20 m plots replicated thrice) were directly seeded in July 1993. Seedling was repeated in July 1994. Tree species seeded were : *Acacia catechu, Acacia nilotica, Azadirachta indica, Dalbergia sissoo, Leucaena leucocphala, Madhuca indica, Pongamia pinnata, Terminalia arjuna, Syzygium cumini*, and *Zizyphus jujuba*, Leguminous forb, *Stylosanths hamata*, and the grass, *Pennisetum pedicellatum* were also seeded in all plots.

Same tree species, as used in direct seeding on flat surface, were seeded on mine spoil slopes in 20 m vertical strips in July 1993 and was repeated in July 1994. Total area seeded was 3 ha. leguminous forb and grass was also seeded as on flat surfaced area.

The growth performance data for flat surface and for slopy area after 3 yrs of seeding are given in Table 4. Growth performance data indicates that legumes were more successful than non-legumes on both sites.

Table 4. Growth performance of tree species 3 yrs after direct seedling on slope of coal mine spoil at Jayant Project, NCL, Singrauli (Based on Singh, *et al*., 1996).

	Legumes			Non-legumes		
	Min.	Max.	Mean	Min.	Max.	Mean
Height (m)	1.71	2.33	2.0	0.92	2.61	1.7
Diameter (cm)	2.03	5.52	3.4	2.28	3.94	3.2
Height:Diameter ratio	31	115	74	30	83	54
Tree volume (cm^3)	1291	9602	5370	478	2573	1749

Thus, direct seeding is a potentially useful tool for revegetation of mine spoils, however, seeding will have to be repeated each year for 3-4 years. The site will then be able to select for seeds having low soil water adaptational syndrome.

Rainwater harvesting tank

Water is a a major limiting factor for revegetation of mine spoils. For rapid, reliable and effective revegetation, availability of irrigation is essential. Therefore, a demonstration experiment was performed to test the feasibility of constructing the tank along the steep slopes of spoils to store enough run off in tank and to grow fruit trees around tank and use of water in life saving irrigation for planted seedlings on slopes.

A tank of 40 m x 40 m size has been constructed at Jayant Project on the base of mine spoils slopes. Around the tank *Emblica officinalis, Zizyphus jujuba, Morus alba, Anacardium occidentale, Mangifera indica. Syzygium cumini, Madhuca indica, Azadirachta indica, Ficus benghalensis, Ficus religiosa,* and *Psidium guajava* (fruit trees and keystone species) have been planted.

Among the seven fruit tree species around water harvesting

tank, for which the data were collected, the maximum growth performance was shown by *E. officinalis* in terms of both height and diameter growth followed by *Zizyphus jujuba*. Whereas *Anacardium occidentale* and *Psidium gaujava* showed minimum growth performance (Singh *et al*., 1996).

Effect of plantations of selected native and exotic species on spoil

For the evaluation of impact of young plantations (4-5 yr old) of native and exotic species on physico-chemical and biological properties of mine spoils at Jayant Project following platn species were selected:

Native legumes:

Albizia lebbec, Albizia procera, Dalbergia sissoo, Pongamia pinnata.

Native nonlegumes:

Azadirachta indica, Dendrocalamus strictus, Shorea robusta, Tectona grandis.

Exotic legume:

Acacia auriculiformis, Casuarina equisetifolia, Cassia siamea

Exotic nonlegume: *Eucalyptus hybrid, Grevillea pteridifolia*,

Compared to native non-leguminous species plots, native leguminous species plots showed greater organic C, nitrate - N, ammonium - N, N-mineralization rate, and microbial C, N and P contents, Except for organic C no marked difference were found between exotic legumes and non-legumes plots. These findings suggested that native legumes are more efficient in bringing out differences in soil properties than exotic legumes in short term (Singh *et al*., 1995).

REFERENCES

Ae, N.J., H. Arihara, K. Okada, T. Yoshihara and C. Johansen, 1990, Phosphorus uptake by pigeon pea and its role in cropping systems of the Indian subcontienent. *Science*, 248: 477-480.

Agrawa, M., J. Singh, A.K. Jha and J.S. Singh, 1993, Coal- based environmental problems in a low-rainfall tropical region, pp. 27- 57. In: R.F. Keefer and K.S. Sajwan (eds.) *Tree Elements in Coal and Coal Combustion Residues*. Lewis Publishers, Boca Raton, USA.

Ashby, W.C., 1987. Forests. pp. 89-108. In: W.R. Jordan III, M.E. Gilpin and J.D. Aber (eds.) *Restroration Ecology*, Cambridge University Press, Cambridge, England.

Bradshwa, A., 1997. Restoration of mined lands - unsing natural processes. *Ecological Engineering*, 8: 225-269.

Bradshaw, A.D. 1983. The reconstruction of Ecosystems. *Journal of Applied Ecology*, 20: 1-17.

Bradshwa, A.D., 1987. The reclamation of derilict land and the ecology of ecosystems. pp. 53-74. In : W.R. Jordan III, M.E. Ciplin and J.D. Aber (eds). *Restoration Eclogy*, Cambridge University Press, Cambridge, England.

Ewel, J.J., 1987. Restoration is the ultimate test of ecological theory. pp. 31-33. In: W.R. Jordan III, M.E. Gilpin and J.D. Aber (eds.) *Restoration Ecology*, Cambridge University Press, Cambridge, England.

Gupta, R.K. and D.L.N, Rao, 1994. Potential of wastelands for sequestering carbon by reforestation. *Current Science* ,66: 378- 380.

Holdgate, M.W. and M.J. Woodman (eds.) 1986. *The Breakdown and Restoraton of Ecosystems*. Plenum, New York, USA.

Jansoon, B. 1997. Preface *Ecological Engineering*, 8: v-vi.

Jankinson, D.S. and J.N. Ladd. 1981. Microbial biomass in Soil: measurement and turnover. pp. 415-471. In: E.A. Paul and J.N. Ladd (eds.) *Soil Biochemistry*. vol. 5. Marcel Dekker, Inc. New York.

Jha A.K. 1989. A note on the root development in dry tropical naturally revegetated coalmine spoils. *Vegetatio*, 85: 67-70.

Jha, A.K. 1992. Evaluation of coalmine spoil as a medium for platn growth in a dry tropical environemtn, India. *Indian Forester*, 118(12): 909-916.

Jha A.K. and J.S. Singh, 1990a. Revegetation of mine spoils: Review and a case study, pp. 300-326. In: B.B. Dhar (ed.). *Environmental management of Mining Operaton*. Ashish Publishing House, New Delhi.

Jha, A.K. and J.S. Singh, 1990b. Vascular flora of naturally revegetated coalmine spoils in a dry tropical environement. *Journal of Tropical Forestry*, 6: 131-142.

Jha, A.K. and J.S. Singh. 1991. Spoil characteristics and Vegetation development of an age series of mine spoils in a dry tropical environment. *Vegetatio*, 97: 63-73.

Jha, A.K. and J.S. singh, 1992. Influence of microsites on redevelopment of vegetaion on coalmine spoils in a dry tropical environemtn. *Journal of Environmental Management*, 36: 95-116.

Jha, A.K. and J.S. Singh,1993a. Growth performance of Certain directly seeded plants on mine spoil in a dry tropical environemtn, India. *Indian Forester*, 117 (II): 920-927.

Jha, A.K. and J.S. Singh. 1993b. Rehabitation of mine spoils, pp. 210-254. In: J.S. Singh (ed.) *Restoration of Degraded Land: Concepts and Strategies*. Restogi Publicatons, Meerut.

Jha, A K. and J.S. Singh. 1994a. Rehabilitation of mine spoils with particular reference to multipurpose trees. pp. 237-249. In: Punjab Singh, P.S. Pathak and M.M. Roy (eds.) *Agroforestry systems for sustainable Landuse*. Oxford and IBH Publishing Co. Pvt. Ltd., New Delhi.

Jha, A.K. and J. Singh. 1994b. Restoration of mine spoils: Concepts and Strategies. pp. 112-118. In: Green Era in Indian Steel: *A Technical Compendium*. Environment Management Division, Steel AUthority of India Limited, Calcutta.

Kline, V.M. and E.A. Howell. 1987. Prairies. pp. 75-83. In: W.R. Jordan III, M.E. Gilpin and J.D. Aber (eds.) *Restoration Ecology*, Cambridge University Press, Cambridge, England.

Lubchenco, J., A.M. Olson, L.B. Brubaker, S.R. Carpenter, M.M. Holland, S.P. Hubell, S.A. Levin, J.A. Mac Mahon, P.A. Matson, J.M. Melillo, H.A. Money, C.H. Peterson, H.R. Pulliam, L.A Real, P.J. Regal and P.G. Risser. 1991. The sustainalbe biosphere initiative: An ecological research agenda. *Ecology*. 72: 371-412.

Powlson, D.S., P.C. Brookes and B.T. Christensen. 1987. Measurement of soil microbial biomass provided and early indication of changes in total soil organic matter due to straw incorporation. *Soil biology and Biochemistry*, 19: 159-164.

Singh, A., A.K. Jha and J.S. Singh. 1996. Influence of NPK fertilization on biomass production of *Pennisetum pedicellatum* seeded on coalmine spoil. *Tropical Ecology*, **37**(2): 285-287.

Singh, J.S. and A.K. Jha. 1987. Ecological aspects of reclamation and revegetation of coalmien spoils. pp. 73-86. In: B.B. Dhar (ed.). *Environemental Management of Mining Operations in India* - A Status Paper. Department of mining Engineering, Banaras Hindu University, Varanashi.

Singh, J.S. and A.K. Jha, 1993. Restoration of degraded land: An overview. pp. 1-9. In: J.S. Singh (ed.) *Restoration of Degraded Land: Concepts and Strategies*. Rastogi Publications, Merrut.

Singh, J.S., A.S. Raghubanshi, R.S. Singh and S.C. Srivastava. 1989. Microbial biomass acts as a source of plant nutrients in dry tropical forest and savanna. *Nature* (London) 338: 499-500.

Singh, J.S., K.P. Singh and A.K. Jha (eds.) 1995. *An Integrated Ecological study on Revegetation of Mine spoil: Concepts and Research Highlights*. An interim Report of a Science and Technology Project sponsored by the Ministry of Coal, Government of India, New Delhi through CMPDI, Ranchi. Department of Botany, BHU Varanasi, 68p.

Singh, J.S., K.P. Singh and A.K. Jha (eds.) 1996. *An Integrated Ecological study on Revegetation of Mine spoil*. Final Technical Report of a Science and Technology Project sponsored by the Ministry of Coal, Government of India, New Delhi through CMPDI, Ranchi, Department of Botany, BHU, Varanasi, 116 p.

Srivastava, S.C. and J.S. Singh. 1988. Carbon and phosphorus in the soil biomass of some tropical soils of India. *Soil Biology and Biochemistry*, 20: 743-747.

Srivastava, S.C., A.K. Jha and J.S. Singh. 1989. Changes with time in soil biomass C,N and P of mine spoils in a dry tropical environment. *Canadian Journal of Soil Science*, 69 : 849-855.

CHAPTER - 7

VAM COLONIZATION AND GROWTH OF *ACACIA MELANOXYLON* IN OPEN CAST COAL MINE DISTURBED SOILS

M.S. Rao, P. Pavan Kumar & S.R. Reddy
Microbiology Section, Department of Botany
Kakatiya University, Warangal - 506 009

INTRODUCTION

Daft and Hacskaylo (1977) reported success in planting of tree species inoculated with VAM fungi in critical sites. Colonization of transplants by VAM fungi has helped in establishment of various plants species on coal mines soils (Reeves *et al.,* 1979). The results of most of the workers on VAM fungi have shown that most tree species respond better with indigenous ones as compared to those isolated from widely separated habitat and genera of unrelated tree species. Revegetation of lands disturbed by mining is reported to have been facilitated by the use of mycorrhizal transplants (Lindemann *et al.,* 1984). Among the different types of approaches for harvesting the beneficial effects of VAM mycorrhizae, inoculation of seed, seed beds and nurseries is a practically feasible proposition as it requires application of small amounts of inoculum for tailoring a large number of seedlings with mycorrhizae in a short space of time. The quality seedlings fortified with selected mycorrhizal fungi have better adaptibility to critical sites because of their tolerance to harsh conditions and their ability to absorb relatively immobile elements such as P, Cu, Zn which are not readily available. Such seedlings are likely to be more persistent and less effected by P immobilization in adverse sites than non mycorrhizal seedlings. In view of the importance of seedling inoculations at nursery level, in the present investigations an attempt has been made to raise the mycorrhizal seedlings of *Acacia melanoxylon* a potential tree species for afforestation with the help of native undisturbed topsoil.

METHODOLOGY

The present investigations were carried out at Singareni Coal mines, Godavarikhani. Mycorrhizal status of undisturbed soils around open cast coal mines was determined by enumerating mycorrhizal colonization of native local plants and also by quantitative and qualitative estimating the mycorrhizal resting spores in rhizosphere. To determine the percentage of VAM infection in selected plant species atleast three plants of the same age from different corners of the field are selected and exacavated. Their feeder roots were collected, washed and immediately preserved in FAA. Phillips and Hayman (1970) procedure was employed to clear and stain the roots. Percentage of infection was calculated by the formula given by Giovanetti and Mosse (1980). Resting spores of VAM fungi were extracted by wet sieving and decanting method of Gerdeman and Nicolson (1963) and subsequently identified by following the key provided by Schenck and Perez (1987). Nursery experiments were conducted at the coal mine site, Godavarikhani. The soils in different combinations as shown in Table-3 were prepared in polythene bags (6 x 12 cm). Previously boiled and cooled seeds of *Acacia melanoxylon* were shown at the rate of five seeds for each bag. After germination, the seedlings were thinned down to one for each bag. Nursery beds were regularly watered and all the nursery maintenance precautions were taken. At the end of 90 days, the saplings were taken out and various parameters were estimated.

RESULTS AND DISCUSSION

A critical perusal of the Table-1 reveals that the native topsoil of the coal mines harboured sufficient quantity of VA mycorrhizal fungi in association with different angiospermic plants. It is also evident that all the host plants growing in the native soil were colonized by VA mycorrhizal fungi. However, the total percentage of infection varied with the host plant. Maximum colonization was observed in *Morus alba* followed by *Zizyphus mauritiana* and *Acacia arabica*. Least infection was observed in *Butea monosperma* and

Waltheria indica. Variation in the mycorrhizal infection in plants belonging to same family and genus was observed by earlier workers (Mayer, 1973. Rachel, 1991, Ravi *et al.*, 1995, Shailaja, 1997). Based on the survey conducted on 2930 species representing 29 families Newman and Reddell (1987) concluded that a large difference exists between the families in the percentage of mycorrhizal colonization.

All the plants species under investigation have shown all the three types of infections viz., arbuscular, vesicular and hyphal. However, *Alternanthera sessilis* is an exception which has not shown any arbuscular infection. Hirrel *et al.*, (1978) suggested that absence of arbuscules could indicate a non-symbiotic colonization. In general, hyphal colonization was found to be more than the other two types of colonizations. The number of resting spores in the rhizosphere soil varied with the host plants. Maximum number of spores were observed in *Zizyphus mauritiana* followed by *Acacia arabica*. Least number of spores were observed in rhizosphere of *Waltheria indica*. No direct correlation was observed between the percentage of infection and resting spore population in rhizosphere. For instance *Morus alba* with 100 percent infection has supported very less number of resting spores.

The present investigations also reveal that the native soils of coal mine harboured all the known six genera of VAM fungi (Table-2). *Glomus* species were found to be associated with all the investigated plants. Similar to present observations, various workers (Gerdeman, 1968; Hayman, 1982; Powell and Bagyaraj and Jha *et al.,* 1994) reported the obiquitous nature of *Glomus* species. Next to *Glomus*, species of *Sclerocystis* followed by *Acaulospora* were found to be associated with large number of hosts. *Gigaspora* species were found in association with 9 host plants only *Scutellospora* and *Entrophospora* were less dominant as they are found to be associated with very few host species. Majority of the host species under investigation harboured more than one genus of the VAM fungi. Intrestingly *Morus alba* with heaviest colonization harboured only two species of the VAM fungi.

Results obtained with experiments on *Acacia melanoxylon* seedlings grown with topsoil (TPS), Open Cast Soil (OCS) are presented in table-3. From these investigations it is evident that the seedlings grown in OCS have shown very poor growth, due to non availability of mycorrhizae and rhizobia for infection and adverse soil conditions. The non-availability of the mycorrhizae, rhizobia and other beneficial micro-organisms can be attributed to the fact that OCS is unearthed from very deep layers of the soil where most

Table 1. Occurrence of VAM fungi in undisturbed soils around the Singareni open cast coal mine in association with different angiospermic hosts

S.No.	Plant species	Family	Vesicular Colonization	Arbuscular Colonization	Hyphal infection	Total	Spore number/ gram of air dried soil
1.	*Acacia arabica*	Mimosaceae	24.4±4.0	35.3±7.0	38.4±9.9	98.12±0.9	55.5±13.1
2.	*Alternanthera sessilis*	Amaranthaceae	16.2±5.7	0.0±0.0	33.5±10.0	49.6±15.6	40.2±12.0
3.	*Azadirachta indica*	Meliaceae	30.6±2.9	34.3±4.6	25.9±4.0	90.9±11.5	39.5±12.3
4.	*Butea monosperma*	Fabaceae	2.5±1.2	12.4±2.0	11.8±4.1	26.7±7.3	24.0±6.9
5.	*Commelina obliqua*	Commelinaceae	26.9±2.9	11.9±2.9	28.9±6.9	67.7±12.7	40.5±9.9
6.	*Gliricidia maculata*	Fabaceae	22.1±1.8	30.2±2.5	32.4±4.0	84.8±8.3	45.2±4.6
7.	*Jatropha heterophylla*	Euphorbiaceae	30.6±16.5	26.9±13.0	32.9±2.9	[illegible]	36.9±20.6
8.	*Lantana camara*	Verbenaceae	30.0±5.6	17.9±2.9	30.9±3.8	78.8±12.3	34.0±11.8
9.	*Morus alba*	Moraceae	22.8±12.6	42.6±10.6	34.6±3.8	100.0±29.1	29.0±12.6
10.	*Parkinsonia aculeata*	Caesalpiniaceae	15.3±3.0	28.3±4.0	29.3±4.3	73.0±11.2	34.3±4.2
11.	*Parthenium hysterophorus*	Asteraceae	19.9±3.4	15.6±3.6	19.9±3.8	55.4±10.9	49.0±23.1
12.	*Peltophorum pterocarpus*	Caesalpiniaceae	22.3±5.1	31.5±3.8	30.2±3.7	84.0±12.6	44.4±8.5
13.	*Sida acuta*	Malvaceae	26.9±7.0	29.9±4.9	33.9±4.4	90.7±16.2	43.9±6.9
14.	*Tribulus terrestris*	Zygophyllaceae	23.0±4.6	19.0±4.9	32.9±4.9	74.8±14.3	34.9±12.0
15.	*Turnera ulmifolia*	Turnaraceae	22.9±4.8	11.9±4.2	29.8±3.0	64.6±11.9	32.9±2.9
16.	*Waltheria indica*	Sterculiaceae	7.9±2.4	1.0±1.0	30.3±4.9	39.3±8.3	14.6±8.7
17.	*Zizyphus mauritiana*	Rhamanaceae	30.8±4.5	29.7±2.2	38.0±1.7	98.4±8.3	60.1±22.2

Table 2. Occurrence of VAM fungi in undisturbed soils around the Singareni open cast coal mine in association with different angiospermic hosts

S.No.	Plant Species	Family	GL	GS	AC	SC	SS	ES
1.	*Acacia arabica*	Mimosaceae	+	+	+	+	–	–
2.	*Alternanthera sessilis*	Amaranthaceae	+	+	+	–	–	–
3.	*Azadirachta indica*	Meliaceae	+	+	–	+	–	+
4.	*Butea monosperma*	Fabaceae	+	+	–	+	+	–
5.	*Commelina obliqua*	Commelinaceae	+	+	+	+	–	–
6.	*Gliricidia maculata*	Fabaceae	+	–	+	+	+	–
7.	*Jatropha heterophylla*	Euphorbiaceae	+	–	+	+	–	–
8.	*Lantana camara*	Verbenaceae	+	+	+	–	–	–
9.	*Morus alba*	Moraceae	+	–	+	–	–	–
10.	*Parkinsonia aculeata*	Caesalpiniaceae	+	–	+	+	–	–
11.	*Parthenium hysterophorus*	Asteraceae	+	–	+	+	+	–
12.	*Peltophorum pterocarpus*	Caesalpiniaceae	+	–	+	+	–	+
13.	*Sida acuta*	Malvaceae	+	+	–	+	–	–
14.	*Tribulus terrestris*	Zygophyllaceae	+	–	+	+	+	–
15.	*Turnera ulmifolia*	Turnaraceae	+	+	–	+	+	–
16.	*Waltheria indica*	Sterculiaceae	+	–	+	+	–	–
17.	*Zizyphus mauritiana*	Rhamanaceae	+	+	–	+	–	+

GL = *Glomus*; GS = *Gigaspora*; AC = *Acaulospora*; SC = *Sclerocystis*;
SS = *Scutellospora*; ES = *Entrophospora*

of micro organisms including the mycorrhiza do not exist. On the other hand, the seedling grown in undisturbed topsoil have shown substantial quantity of mycorrhizal infection and nodulation. Consequently, the seedling growth was also profuse. OCS in combination with various proportions of TPS has exhibited dramatic effect on the seedling growth when compared with the whole OCS. Mycorrhizae infection, spore number, nodulation and seedling growth increased. Mycorrhizal seedlings are ideal for outplanting on the dumps of open cast mines. Because of the problems associated with mass production of mycorrhizal inoculum for field application, nursery grown mycorrhizal seedlings are ideal for outplanting. The present investigations indicate that the undisturbed TPS soil alone or in mixture with OCS can be used at nursery level for raising the nursery seedlings.

Table 3. VAM colonization, nodulation and growth of *A. melanoxylon* seedlings in undisturbed top soil and open cast coal mine soil.

1.	Top soil	100%	78	282	28	0.9	0.18	18.6	46.5	0.72	3.41
2.	TPS 75% + OCS	25%	56	203	20	0.6	0.11	16.7	38.1	0.54	2.56
3.	TPS 50% + OCS	50%	44	156	17	0.5	0.09	15.8	31.9	0.38	2.12
4.	TPS 25% + OCS	75%	35	109	13	0.5	0.07	13.2	22.3	0.26	1.61
5.	Open cast soil	100%	-	-	-	-	-	9.8	15.6	0.18	1.02

TPS = Top Soil; OCS = Open cast soil.

REFERENCES

Daft M.J., and E. Hacskaylo 1977. Growth of endomycorrhizal and non-mycorrhizal red maple seedlings in sand and anthracite soil. *Forest Sci.* 23 : 297-316.

Gerdeman, J.W. and T.H. Nicolson 1963. Spores of mycorrhizal endogone species extracted from soil by wet sieving and decanting. Trans. Br. Mycol. Soc. 46 : 235-244.

Gerdeman, J.W. 1968. Vesicular-arbuscular mycorrhizae and plant growth. *Ann. Rev. Phytopathol.* 6 : 397-418.

Giovannetti, M. and B. Mosse 1980. An evaluation of techniques for measuring vesicular arbuscular mycorrhizal infection in roots. *New Phytol.* 84 : 489-500.

Hayman, D.S. 1982. Influence of soils and fertility and survival of vesicular-arbuscular mycorrhizae. *Phytopathology*. 72 : 1119-1125.

Hirrel, M.C., H. Mehravaran and J.W. Gerdemann 1978. Vesicular arbuscular mycorrhizae in the Chenopodiacea and Cruciferae do they occur ? *Can. J. Bot.* 56 : 2813-2817.

Jha, D.K., G.D. Sharma and R.R. Mishra 1994. Ecology of vesicular arbuscular mycorrhiza. Microbes and Environment pp. 199-208 (Eds. A.B. Prasad and R.S. Bilgrami), New Delhi.

Lindemann, W.C., Lindsey, D.L. and Fresquez, P.R. 1984. Amendment of mine spoil to increase the number and activity of micro organisms. *Soil. Soc. Am. J.* 48(3) : 574-578.

Mayer, F.H. 1973. Distribution of ectomycorrhiza in native and man-made forests. In ectomycorrhizae. Their ecology and physiology. (Eds. G.C. Marks and T.T. Kozlocoski) Academic Press, New York, pp. 79-105.

Newman, E.I. and P. Reddell 1987. The distribution of mycorrhizas among families of vesicular plants. *New Phytol.* 106 : 745-751.

Phillips, J.M. and D.S. Hayman 1970. Improved procedure for clearing roots and staining parasitic and vesicular arbuscular mycorrhizal fungi for rapid assessment of infection. *Trans. Br. Mycol. Soc.* 55 : 158-160.

Powell, C.L.L. and D.J. Bagyaraj 1984. Field inoculation with VA mycorrhizal fungi. In VA mycorrhizae. Florida : CRC press, Boca Raton, pp. 205-222.

Rachel, E.K. 1991. Studies on vesicular arbuscular fungi. Ph.D. Thesis Kakatiya University.

Ravi, K.B., J. Prabhakaran and S. Mariappan 1995. Survey of vesicular arbuscular mycorrhizae in agro forestry tree in alfisol. In mycorrhizae : Biofertilizers for the future. Proc. 3rd Nat. Con. in mycorrhiza (Ed. Alok Adholeya and Sujan Singh) 95-99.

Reeves, F.B., D. Wagner., T. Moorman and J. Kiel 1979. The role of endomycorrhizas in revegetation practices in the semi-arid west. I. A comparison of incidence of mycorrhizae in severely disturved and natural environments. *Amer. J. Bot.* 66 : 6-13.

Schenck, N.C. and Y. Perez 1987. Manual for the identification of VA mycorrhizal fungi. In VAM Florida University. Gainesville, U.S.A., 245 pp.

Shailaja, 1997. Studies on vesicular arbuscular mycorrhizal fungi with special reference to *Sesamum indicum* L. Ph.D. Thesis Kakatiya University.

CHAPTER - 8

BIOLOGICAL RECLAMATION OF COAL MINING OVERBURDEN DUMPS

R.S. Singh, S.K. Chaulya & S.K. Singh

Central Mining Research Institute
Barwa Road, Dhanbad - 826 001

INTRODUCTION

Ecosystem destruction by mining is an inevitable part of industrialization and civilization. Large scale mining operations have the potentials to destroy the flora, fauna and contaminate the soil, air and water in mining regions (Davcheva, 1990). Surface mining operations, which often reslult in removal of huge quantities of overburden dumps in India has been of great concern to both the public and government agencies. Mine spoils present very rigorous conditions for soil microbe and plant growth, because of their low organic matter content and unfavourable soil chemistry, poor structure either course or compact, and high isolation (Mayer, 1973, Jha & Singh, 1994).

Rehabilitaion of overburden waste dumps is thus a complicated problem that requires a thorough understanding of ecological principles. The first concern is to maximize ecological preventive measures such as ensuring waste dump stability, control of toxic mine drainage and sedimentation in water courses. Restoration of bio-diverse habitat is required to help in the proper functioning of the ecosystems a measure which needs to be considered during the planning stage of operations. So, there is a need to optimize the degraded environment both for individual species and for the entire ecosystem, because both have specific physical and chemical requirements. (Bradshaw, 1996).

The paper highlights these preventive and restorative management aspects of open pit mining in India together with the results of a project at the Jharia coalfields which is one of the oldest mining areas in the country and is an dire need of rehabilitation.

Importance of Pilot Projects

The planning procedures, implementation and monitoring required for rehabilaitation of open cast coal mines are known, but it has not been a common practice throughout India till recent years, because most of the mining companies have little direct experience with rehabilitation planning. Therefore, pilot projects are needed for each site to demostrate potentiality and sustainability for rehabilitation. Each pilot project can serve as a model for the region. The following components need to be included in each demonstrative project:

(i) **Goals and objectives :** These may be embodied in a master plan or may be simply a statement of priorities and general policy at the inception stage. It should be based on sound technical analysis and prepared by a qualified body.

(ii) **Technical Plan :** It should reflect in detail, how the objectives will be achieved, including operation plans and completion criteria. The plan must include site selection, material movement, protection and rehabilitation of the resources of the site. The plans should take into account the type, degree and orientation of slope, location of high and low points, size and location of drainage, soil requitements and the most appropritate post-mining land use, such as native vegetation restoration for forage and browse for wildlife, or pasture and feed for livestock.

(iii) **Decision Making Process :** A decision making process needs to be established which allows for adaptability of existing plans and processes, based on the interest of the mass people widest representation of interests.

(iv) **Vegetation Studies :** Pilot projects need to involve both laboratory and field trial experiments. Laboratory or pot testing includes small-scale vegetaion tests on the waste in either its natural or modified state. Once success is obtained in the laboratory, field trail should begin.

STUDY SITE AND METHODS

Location and Topography

The study sites are located in the Katras and Mugma areas of Bharat Coking Coal Limited, and Coal India Limited respectively (Fig.1).

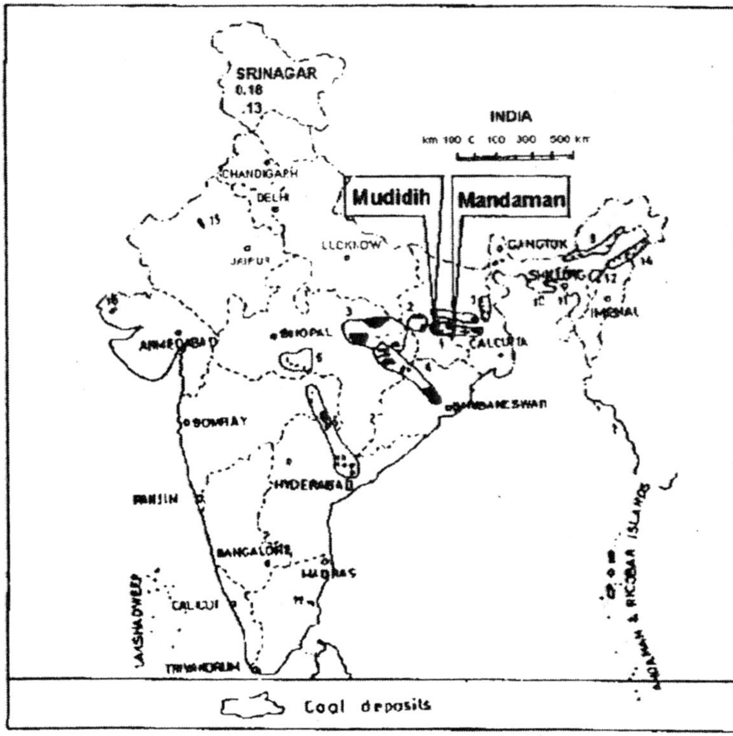

Fig. 1. Location map of the study sites

Huge quantities of overburden materials were found scatteredly dumped at this site. 'Mean Reduced Level (R.L.) of Mandaman and Modidih dumps' were 151.14 m and 226.00 m; the height varied from 14-29.5 m and 20-30 m and slope of the dumps ranged form 31-33 and 33-35 respectively (Figs. 2 & 3). Mining was carried out by opencasting, and the present extraction is continuing by shovel- dumper combination. The selected dump for pilot project were six years and twelve years old with an area of 22500 m^2 for Mandaman and Modidih sites, respectively.

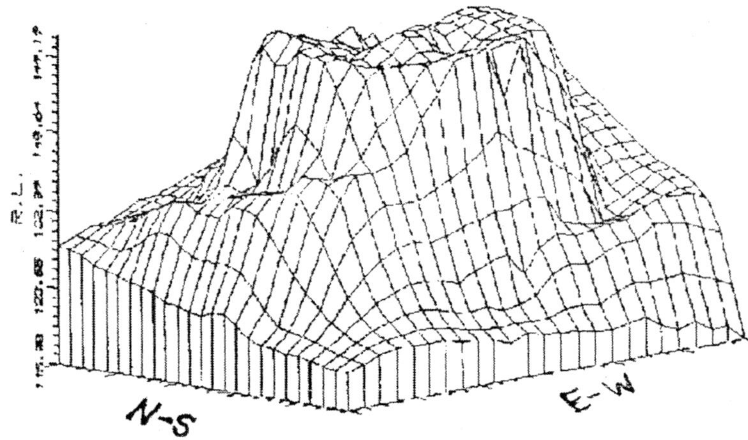

Fig. 2. 3D View of Mandaman Dump

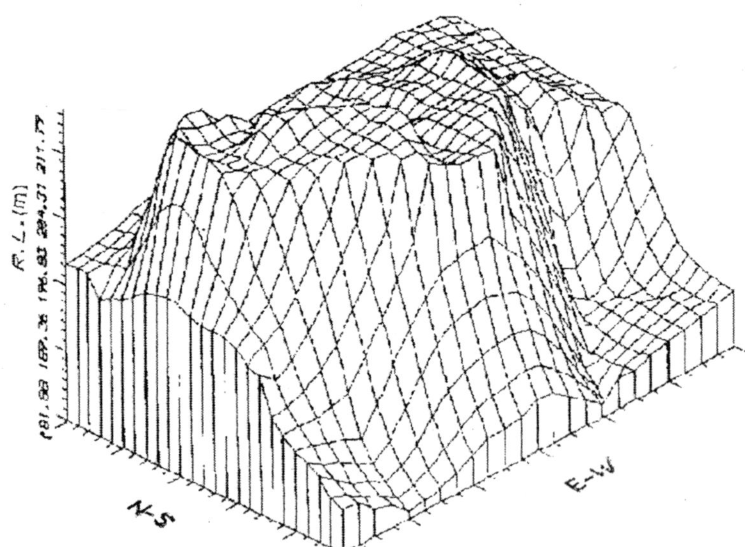

Fig. 3. 3D View of Modidih Dump

Climate

The climate is dry tropical and the year can be divided into cold winter (December to February), a very hot summer (April to June) and rainy season (July to September). March and Ocotober constitute the intermediate months between winter and summer season and rainy and winter season, respectively. Mean daily temperature within an annual cycle rages from 10-28°C and mean daily maximum 26-45°C. The average annual rainfall is 1376 mm, of which 1107 mm occurs between late June to September.

Geology and Soil

The bedrock was formed of medium to coarse grained sandstone clay with ferrugineous bands the carbonaceous shales. The soil surface layer was 10-11 cm thich, gray brown to very pale brown in colour, sandy loam to clay loam in texture with subangular blocky structure. Ferromagnetic concentraions and clay content were found in sub soil. The overburden consisted of alluvium loose and sand, gravel, shale and sandstone (Hussain *et. al.*, 1987).

Vegetation

The local vegetation is a typically mixed dry deciduous forest with domination of sal, shisam and palas forest. On mine spoil the most dominant and frequent plant species are invader shrub species viz. *Leonotis nepatifolia, Eupatorium odoratum and Lantana camera*. These species are causing nuisance by preventing the regeneration of native species in the area. Consequently the whole ore is occupied by these weed species. Major grass species are *Saccharum spontaneum and Eragrostis tenella* (Singh *et al*., 1994).

METHODS

Healthy seeds of seven tree species were sown in 1.5 cm depth in poly-ethylene bags in the first week of May, 1993 at CMRI laboratory. Saplings and grass tiller were transported to the overburden (OB) dump site. Length and diameter were measured after transplanting the plants in OB dump in the month of July (1993) and subsequently followed in 1994 and 1995. While healthy tillers of three grass species were also sown on the slopes of overburden dumps. The study was based on laboratory experiments and two years field trial experiment.

Following plant species were selected for pilot project study (local names in parenthesis) :

Tree Species

(a) *Acacia nilotica* (Babool)

(b) *Albizia procera* (Siris)

(c) *Azadirachta indica* (Neem)

(d) *Dalbergia sissoo* (Shisam)

(e) *Delonix regia* (Gulmohar)

(f) *Leucaena leucocephala* (Subabool)

(g) *Pongamia pinnata* (Karanj)

Grass Species

(a) *Cymbopogon citratus* (Lemon grass)

(b) *Saccharum spontaneum* (Kashi grass)

(c) *Vetiveria zizanioides* (Khas grass)

RESULTS AND DISCUSSION

Laboratory Experiment

In laboratory Seeds of Shisham, Neem, Siris, Gulmohar, Babool, Subabool, and Karanj were sown to quantify the germination potential of the tree species on OB dumps mine spoilunder the lab. condition. All the tree seeds exhibited more than ninety per cent (90%) germination potential. In pot experiment maximum values for shoot length observed for Shisham and minimum for Babool. When root biomass was estimated Shisham tree has shown maximum root biomass along with several root nodules. While in grass species native Kashi grass exhibited maximum growth followed by Khas and Lemon grass.

Field Experiment

Average survival rate of tree species was 57% after two years across all the sites. Pilot project observation has indicated 280 cm increase in shoot length and 46 mm increase in diameter for Shisham tree after two years of plant growth performance on Mandaman OB dump (Fig. 4 & 5). Subabool exhibited maximum growth (256 cm) on Modidih OB dump and its growth was not significantly higher than Shisham tree (245 cm), however, its diameter was lesser than the native tree Shisham species. Our laboratory experiment has also indicated that native tree species (Shisham) was the fastest growing plant species and does not require any extra organic matter inputs.

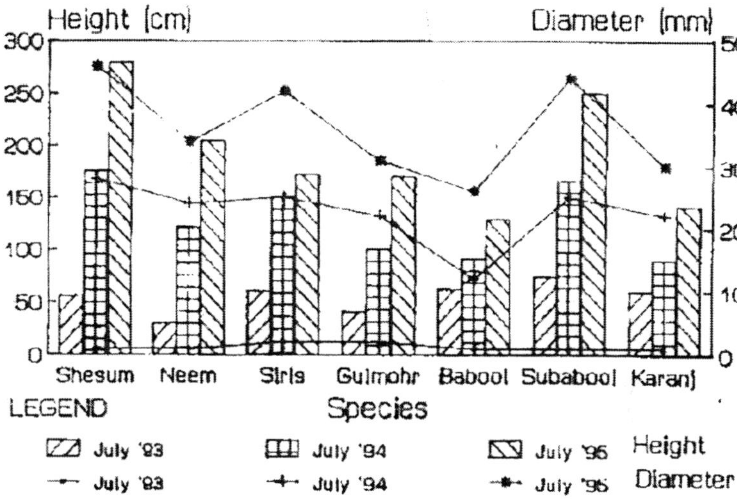

Fig. 4. Growth performance of trees on Mandaman OB Dump

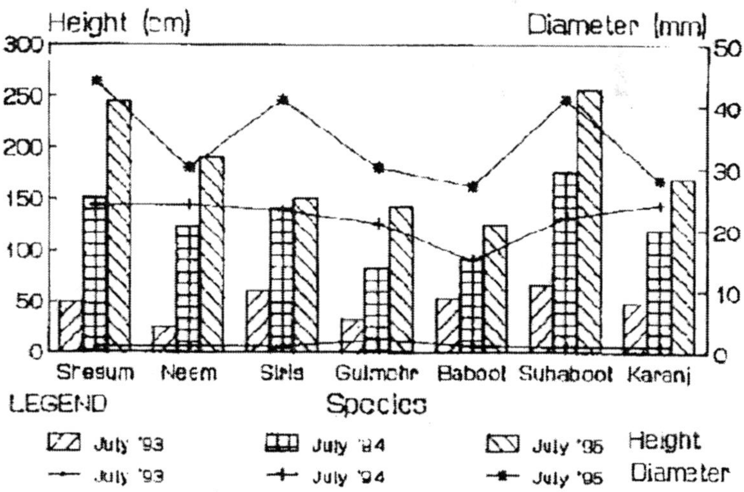

Fig. 5. Growth performance of trees on Modidih OB Dump

Soil analysis indicated that present OB dumps is very impoverished and slightly alkaline in reaction. Therefore, the local plant species have been selected for revegetation purpose. Our study has also indicated that roots of a Shisham tree has shown enormous root nodules containing nitrogen fixing bacteria in all the overburden dump sites. Recently Singh *et al.* (1996) opined that roots of leguminous plant species fix substantial amount of atmospheric nitrogen on the waste dump. Thus, the native perennial species is important for the revegetaion of overburden dump and conservation of local biodiversity.

Among the selected plant species, Shisham and Subabool are the successional tree species with range of tolerance for the growth on nutrient poor mine spoil. Three grass species were planted on the slopes of the overburden to compare their growth performance and the capacity to check soil erosion. The slope of the overburden dump had abundant growth of Kashi grass (Fig. 6). Kashi grass exhibited mamimum growth (232 and 241 cm) with ground spreading capability alongwith high regenerative capacity preventing the soil run-off and slope failure of the dump slope. Minimum growth was

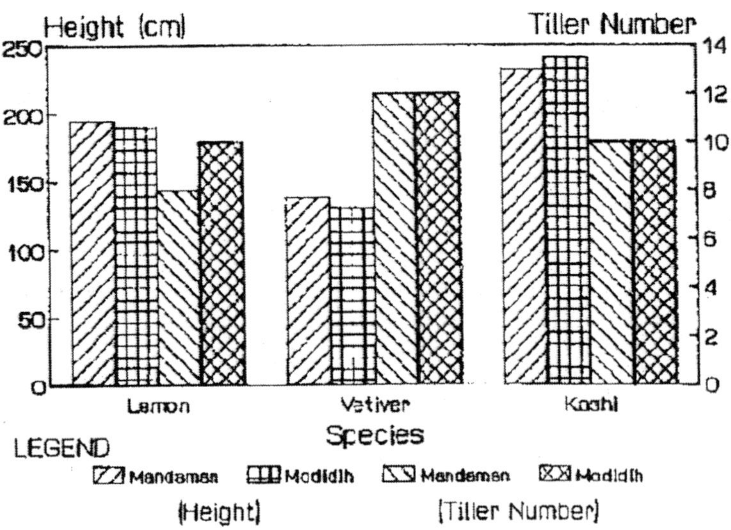

Fig. 6. Growth of Grasses on OB Dump slopes (July 93 – July 95)

observed for Lemon grass. It was also observes that Lemon and Khas grasses do not have self regenerative capacity. Therefore, native Kashi grass species is an appropriate for dump slope stabilisation. Application of farm yard manure (FYM) during the plantation of saplings and grasses and addition of plant residue have great potential to active the microbiological processes. Jha and Singh (1994) stated that use of legumes was more advantageous than fertilisers both in terms of physical and chemical properties of mine soil. Consequently, the whole barren rocks and soils after two years are now converted into a stable overburden dump with beautiful greenery and abundance of livestock productivity along with local wildlife habitation.

ACKNOWLEDGEMENT

Authors are thankful to Director, Central Mining Research Institute for giving permission to present this paper. The Ministry of Coal, Goverment of India, New Delhi is sincerely acknowledged for sponsoring this study.

REFERECNCES

Bradshwa, A.D. 1996. The reclamation of derelict land and the ecology of the ecosystem. 53-74. In W.R. Jordan III, M.E. Gilpin and J.D. Aber (eds.) Restoration Ecology, A Synthetic Approach to Ecological Research, Cambridge Press.

Davcheva, I., N. 1990. Coal mine enterprise - an artificial component in natural ecosystem. In A.K.M. Rainbow (ed.) Proc. of the Third Int. Symp. on Reclamation, Treatment and Utilization of Coal Mining Wastes, Glasgow, pp. 119-124.

Dhar, B.B. 1996. Environmental management system for closure and best practices in Indian mining industry. 6th Workshop of the Mining and Environment Research Network. 4-9 August Harare, Zimbabwe.

Hussain, N., K. Nayan & S. Ali 1987. Report on intensive detailed soil survey for rehabilitation of soil in Jukunda, O.C.P. (BCCL) under environment reclamation programme, Dhanbad, Bihar. All India Soil and Landuse Survey Organisation. Ministry of Agriculture and Rural Development, Government of India. New Delhi.

Jha & Singh, 1994. Restoration of mine spoils : concepts and strategies. Green Era in India Steel - A technical compendium SAIL, Bokaro.

Meyer, F.H. 1973. Distribution of ectomycorrhizae in native and man made forests. In G.C. Marks and T.T. Klozlowski (eds.). Ectomycorrhizae : Their Ecology and Physiology. Academic Press, NY, pp. 79-105.

Singh, R.S., B.K. Tewary & B.B. Dhar 1994. Effect of surface mining on plant biomass and productivity in a part of Dhanbad Coalfield areas. In S.P. Banerjee (ed.) *Second National Seminar on Minerals and Ecology*. 103-109, Oxford & IBH publisher, New Delhi.

Singh, R.S., S.K. Chaulya, B.K. Tewary & B.B. Dhar 1996. Restoration of a coal mine overburden dump - A case study. *Coal International*. 80-83.

CHAPTER - 9

RECLAMATION OF WASTELANDS AND PROTECTION OF ENVIRONMENT : A REVIEW

A. K. Kandya

Deptt. of Botany, Dr. H.S. Gour University, Sagar (M.P.)

INTRODUCTION

Wastelands development has been in the news in India in the recent past with the proposed scrapping of the National Wastelands Development Board (NWDB) which was set up with the purpose of reclaming degraded land and restoring the forest cover. That the government is still keen on wasteland development is evident from the fact that it plans to set up a Technology Mission on wastelands development in the Ministry of Environment and Forests.

With the exploding human and cattle populations in our country, it has become compulsory for us that every inch of land should be made use of, in productive purposes. It has been measured that at present, about 175 million hectares of land in our country is categorised under various kinds of wastelands and every year, about 1.3 million hectares of forest land is coming under the category of wastelands (Anon, 1984).

Definition

Wasteland has been defined in varied ways by different agencies. According to the society for the Promotion of Wastelands Development (SPWD), an NGO in our country "Any land which is not producing green biomass consistently" must be taken as wasteland. But according to the Committee of Wasteland Survey and Reclamation, all barren lands, un-culturable lands and those lands which are not available for cultivation, are wastelands. As per the American Society of Soil Science, a piece of land, not capable of producing materials or services of values, is wasteland. Actually, wasteland is the ruined, devastated, un-used or underused land lying unproductive and useless. More recently, all those lands have been categorised under wastelands which are (1) Ecologically unstable (2) The land which is neither under forests or agriculture, not assigned for specific purposes such as township, road, national parks, mines, sanctuaries industries, or irrigation projects (2) Which has developed toxicity in the root zones of most plants and (4) whose top soil is nearly completely lost (Prasad, 1988).

Table 1. Area under different kinds of wastelands in India

Wasteland category		Area (Million ha.)
1.	Serious water and wind erosion	150.00
2.	Shifting cultivation	3.0
3.	Water-logged area	6.0
4.	Saline soils	4.5
5.	Alkaline soils	2.5
6.	Riverine soils	2.4
7.	Other culturable lands	6.6
	Total	175.0

Types of Wasteland

Whatever is the definition, wastelands can be categorised into 4 types, i.e.

(1) Ecological wastelands which include degraded forests, guillied and ravinous lands, naked hill slopes, saline and alkaline soil, desert and sand dunes.

(2) Developmental wastelands - including mine spoils, over burdons, water-logged areas formed by seepage of canal irrigation etc. Such wastelands result from economical compulsions and changing economic policies.

(3) Culturable wastelands can be reclaimed by certain methods. Gullied and inaccessible ravinous lands, undulating uplands with or without scrubs, surface water-logged lands, marsh lands, salt affected lands and mining spoils belong to this category.

(4) Un-culturable wastelands are those which can not be reclaimed. Barren, rocky and stony lands, steep slopy areas, snow covered and/or glacial areas are included in this category.

Reclamation of Wasteland : Strategies and Success

For the reclamation of wastelands, Govt. of India established a National Wastelands Development Board on the 24th April 1986 under the chairmanship of Dr. M.G.K. Mehon, the then Member of the Planning commission. In Madhya Pradesh (Table-2) total 92.2 lakh hectares of land is categorised as wasteland. Out of it, degraded forests possess the maximum part of 77 lakh hectares.

It has now been realised that afforestation in bare areas and reforestation in degraded forests be done to reclaim the wastelands. And, plantation be done by most modern methods of biotechnology to achieve the maximum success in shortest time, i.e. use of bio-fertilizers, mist propagation of vegetative.

Table 2. Area under different kinds of wastelands in M.P.

Wasteland category		Area (Lakhs ha.)
1.	Degraded forests	77.0
2.	Mined-out areas	1.2
3.	Ravines	7.4
4.	Bhata Lands (Hard Lateritic soil)	3.7
5.	Saline and Alkaline soils	2.4
6.	Water-logged and periodically in undated area	0.7
Total area of wastelands		92.2

Cutting, tissue culture, areial seeding in inaccessible revinous areas and establishment of forest based industires where all raw materials produced can be consumed.

At the very first step, using remote-sensing technique potential sites for afforestation and reforestation are being identified in all parts of the country (Jadhav *et al.*, 1989). Increasing fronteirs of degraded forests are being checked by stall feeding of cattle of a group of villages on a fixed site, so that seedlings do not die due to grazing and trambling and regeneration of forest is not arrested (Prasad and Bhatnagar, 1988). Then, most of the forest tree species have been tested for their suitability to grow in different kinds of wastelands. As we know, different kinds of plants have been categorised according to the soil in which they grow (Table-3).

Table 3. Various plant categories and the soils in which they grow

Plant categories		Grow on soil types
1.	Oxylophytes	Acidic soil
2.	Halophytes	Saline (salty) soils
3.	Psammophytes	Sand
4.	Lithophytes	Rocks
5.	Chasmophytes	Rock crevices

We also know that the soil af. ects the plant in different ways in various phases of its life cycle as detailed below in Table-4.

Table 4. Effects of soil on different phases of plant life.

1.	Ability of seeds to germinate
2.	Erectness of the plant
3.	Vigour of the vegetative parts
4.	Woodiness of the stem
5.	Depth of the root system
6.	Resistance of susceptibility to drought/frost
7.	Date of appearance of flowers
8.	Number of flowers per plant

Actually soils have been categorised into 5 main kinds depending upon the percentage of $Al_2O_3.SiO_2$ (Alumino-silices) called minerals and found in the clay protion of soil (Table-5).

Table 5. Soil types based on the percentage of alumino-silices of minerals

Soil type	% of Al_2O_3. SiO_2
Grey desert soil	3.62
Prarie soil	3.17
Alkali soils	3.01
Podsols	2.84
Lateritic soils	1.20

The chief defects of saline soil are low N_2, low Ca high impermeability, absence of bacterial activity and deficiency of organic matter. These soils require the treatment with gypsum salt

($CaSO_4$) before any plantation is made on them. In this way they are reclaimed

$$\begin{matrix} & Na & \\ Na & Soil & Na \\ & Na & \end{matrix} + CaSO_4 \longrightarrow Na_2SO_4 + \begin{matrix} & Ca & \\ Ca & Soil & Ca \\ & Ca & \end{matrix}$$

Addition of NH_4 salts increases the N_2 content of soil and makes it better for plant growth. Since, different kinds of wastelands possess very specialised types of ecological conditions and nutritional status of soil, different kinds of species grow on them as shown in Table-6.

Methods of Planting

In different kinds of wastelands, different methods of planting are required to be used for different plant species (Table-7). Most species can be planted either by direct sowing of seeds or by polypotted seedlings. But, there are some which grow well by root-cuttings, stumps, rhizomes or bulbils (Agave) and root tufts (*Saccharum munja, Dichanthium annulatum* and other grasses).

Table 6. Tree species suitable for different kinds of wastelands

Kind of wasteland		Tree species suitable
1.	Coal-mine over-burdens	*Encalyptus hydrid, Acacia auriculiformis, Dalbergia sissoo, Albizias and other Acacias etc.*
2.	Bauxite-mine over-burdens	*Eucalyptus camaldulensis Acacia auriculiformis, Gravillea robusta, G. pteridifolia, Pinus caribaea, P. roxburghii*
3.	Saline and Alkaline soil	*Acacia arabica, Butea monosperma, Tamarix aphylla, Azadirechta indica, Albizia lebbeck, A. procera, Pongamia pinnata, Tamarindus indica, Terminalia arjuna, Salvadora persica, Prosopis juliflora etc.*
4.	Dairy effluent area	*Syzizium cuminii*
5.	Sand dunes	*Acacia tortilis, Colophospermum mopane Prosopis cinerarea, Ziziphus rotundifolia, etc.*

6.	Coastal sandy soils	*Casuarina equisetifolia, Calophyllum inophyllum, Borassus flabellifer, Eucalyptus hybrid, Anacardium occidentale, Manilqkara littoralis etc.*
7.	Periodically inundated soils	*Terminalia arjuna, Syzizium cuminii, Trewia nudiflora, Sapium sebiferum etc.*
8.	Limenstone soils	*Bahunia retusa*
9.	Road sides	*Pongamia pinnata, Lagerstroemia speciosa, Cassia siamea, C. fistula, C. glauca, Bauhnias, Jacaranda mimosifolia* etc.

Table 7. Different methods of planting suitable for different plant species

1.	Direct sowing of seeds	*Acacia auriculiformis, A. mellifera, Acacia nilotica, A. tortilis, A. catechu, Anacardium occidentale, Casuarina equisetifolia, Cassia siamea, Holoptelia integrifolia, Leucaena leucocephala, Melia azedarach, Morinda tinctoria, Pongamia pinnata, Tecoma orientalis, Ceiba pentandra, Delonix regia, Dichanthium annulatum.*
2.	Polypotted seedlings	*A. auriculiformis, A. mellifera, Albizia lebbeck, A. procera, A. flacataria, A. amara, Adina cordifloia, Artocarpus hirsuta, Anacardium occidentale, Cassia auriculata, C. nodosa, Gmelina arborea, Holoptelia integrifolia.*
3.	By root cuttings	*Saccharum munja, Eulaliopsis binata, Dichanthium annulatum, Arundodonax.*
4.	By Stumps (Root-shoot)	*Coccus nucifera, Dalbergia sissoo, Melia azedarach, Terminalias, Gmelina.*
5.	By Branch cuttings	*Bouganvillea, Bambusa vulgaris, Salix tetrasperma*
6.	By Rhizome	*Bambusa arundinacea, Dendrocalamus strictus, Arundo donax*
7.	By Root tuffs	*Saccharum munja, Eulaliopsis binata, Dichanthium annulatum*
8.	By Bulbils	*Agave* sp.

Aerial seeding : Of all the kinds of wastelands ravinous areas, arid and semi-arid areas and high rainfall highlands are very large and rather inaccessible. They require aerial dissemination of seeds which is a fast method and broadcasts the seed uniformly. The seeds are sprayed just before the first spell of rainfall so that they can germinate on getting rains. There are different species which are suitable for different kinds of wastelands for the aerial broadcasting of seeds (Table-8). But, before broadcasting, the germination capacity of seeds be tested. Table-9 gives an idea of percentage germination of several tree species which is retained for longer periods. Therefore, those species be selected whose seeds possess high germination capacity (Prasad and Kandya, 1991).

Table 8. Tree species suitable for aerial seeding in different kinds of wastelands

Ravinous Areas	Arid and semi-arid areas	High ranfall highlands
Acacia catechu	*Acacia modesta*	*Acacia benthamii*
Acacia albida	*Acacia mellifera*	*Acacia delbata*
A. camplycantha	*Acacia tortilis*	*A. decurrens*
Ailanthus excelsa	*A. leucophloea*	*A. farnesiana*
Cassia siamea	*Azadirachta indica*	*A. mearnsii*
Dalbergia sissoo	*Boswellia serrata*	*A. melanoxylon*
Dendrocalamus	*Cassia auriculata*	*Artocarpus* sp.
Dichrostachys	*Cassia fistula*	*Bombusa*
Holoptelia	*Prosopis juliflora*	*Cleistanthus*
Madhuca indica	*Prosopis spcigera*	*Aegle marmelos*
Pongamia pinnata	*Sesbania*	*Albizia amara*
Tecomella undulata	*Sterculia* sp.	*Amora wallichi*
Tectona grandis	*Tecomella undulata*	*Diospyros sps.*

Table 9. Germination capacity of seeds of some forestry species (Many more are available)

100-90%	*Artocarpus chaplasha, Acacia catechu, Albizia lebbeck, Anacardium occidentale*
90-70%	*Acer campbelli, Ailanthus grandis. Butea monosperma, Pinus roxburghi, Azadirachta*
70-50%	*Bambusa bambus, Betula alnoides, Terminalia tomentosa, Dendrocalamus strictus*
50-30%	*Cedrus deodara, Salmalia malabarica*
10-5%	*Buxus wallichiana, Cupressus torulosa.*

Grass/Tree species suitable for all the 3 areas : *Acacia auriculiformis, A. nilotica, A. segegal, Gmelina arborea, Cenehrus ciliaris, Dichanthium annulatum, Pennisetum pedicillatum, Stylosanthes.*

Table 10. Rooting response of stecklings of some plant species under mist chambers

Species		% of steckings struck roots	Av. number of root/ steckling	Day required
1.	*Araucaria cunninghammi*	3.0	2.3	30-35
2.	*Azadirachta indica*	32.2	8.0	20-25
3.	*Butea monosperma*	80.0	6.0	20-34
4.	*Casuarina equisetifolia*	50.0	15.0	20-25
5.	*Dalbergia latifolia*	8.0	2.0	20-25
6.	*Dalbergia sissoo*	100.0	6.0	15-25
7.	*Eucalyptus tereticornis*	45.8	3.5	40-60
8.	*Ficus benghalensis*	16.6	6.2	30-40
9.	*Ficus lacor*	65.5	1.2	30-40
10.	*Ficus religiosa*	66.6	4.0	30-40
11.	*Grevillea pteridifolia*	40.5	4.0	10-15
12.	*Lagerstroemia*	58.0	5.0	26-30
13.	*Pongamia pinnata sp.*	72.0	4.0	10-25

Vegetative multiplication of Forestry Trees

For the intensive management of wastelands, it is essential to develop fast and economical methods of raising superior planting stock. Pencil sized diameter and 15 cm long stem cuttings taken from healthy trees of desired species are being used as stecklings for this purpose. Most of the plants have a capacity to develop roots in the stem cuttings if congenial environmental condition are given to them. Usually, such cuttings die due to desiccation because of the fast rate of traspiration from the cut ends. But, if the desiccation is checked, they may not only survive but also give rise to lateral roots. Such a success is obtained in Mist Propagation Chambers where 95 to 100% relative humidity and about 25°C temperature are maintained. The stecklings are inserted in the growing medium which is normally sand or vermiculite. The advantage is that these two media have no water logging. Various concentrations of rooting harmone, i.e. IBA are used to dip the tip of the stekling which is to be inserted in the growing medium. Intermittent spray of mist is given automatically for 1 or 2 minutes after each 15 minutes to keep the atmosphere saturated. Such mist chambers can be made by concrete or Mist Pits and mist beds can be made by using an iron or plastic frame covered by polythene sheet. Plenty of water is supplied to the soil which converts into vapour and keeps the surroundings saturated. Mist pit is of 2m x 1m x 1.5m and sprayed with some fungicude (Kulkarni, 1987).

Only callus formation

The cost of a plantlet is very low (40 to 75 paise) in the mist chamber or poly-chambers made in the field. But, if a concrete chamber is made, the cost becomes higher due to investments in the construction of water storage tank, motor, pipe-line-fittings with several nozzles, humidistat for maintaining the RH more than 70%, some heating arrangement to circulate hot air in the mist chamber during the cold days of winter season etc.

Table 11. Average cost of a plantlet by mist propagation technique

Pongamia pinnata	Rs. 0.50
Butea monosperma	Rs. 0.40
Eucalyptus sp.	Rs. 0.60
Casuarina equisetifolia	Rs. 0.75
Azadirachta indica	Rs. 0.60
Dalbergia sissoo	Rs. 0.40

Most important aspect in the reclamation of wastelands is the use of biofertilizers which are of 3 types, i.e. Dizotrophs (Nitrogen fixers); Phosphate solubilizers and mineralizers; and Nutrient absorbers. There are a large number of advantages from biofertilizers (Table-12).

Table 12. Advantages of biofertilizers

1	Cheaper than conventional nitrogenous and phosphoric fertilizers.
2	Required in lesser amounts i.e. 100 to 1000 g/ha.
3	Improve the productivity of forest by biological conditioning of soil.
4	Helps in soil conservation by maintaining soil health.
5	Helps in better root formation in plants by efficient absorption of water and assimilation of nutrients.
6	Easy to transport
7	Increase the resistance of seedlings against fungal diseases.
8	Shorten the germination period of seeds and induce healthy germination due to growth promoting hormones, metabolites and biologically active substances secreted by inoculants.
9	Minimise the loss of essential elements and promote vigorous vegetative growth.

Biofertilizers are specially used in mined-out over-burdens where large scale open cast mining mechanised operation make vast tracts of land upside down. Their vegetal cover is completely destroyed and fertile layer of soil goes to deeper layers. Fertility of the top layer is almost nil, which does not allow generation of any green cover. Biofertilizers are given to the seeds and fast regeneration is obtained. Table-13 gives the methods of application of biofertilizers.

Tissue culture plants (TCPs)

It has been realised that sometimes lots of seeds fail to germinate due to unfavourable conditions or due to their own low germination capacity. In this situation, if a small seeding or plant-let is planted there are more chances of its continued survival. Tissue culture is a method of producing a very large number of plantlets from a very little amount of any part of a plant. But, through a series of experiments in different tissue culture laboratories in different countries a large number of tree species have been tested and only a little success has been found (Table-14).

Table 13. Methods of application of biofertilizers

1.	Power method	Seeds are sprinkied with 10% sugar solution or 30-40% water soluble gum. Powdered biofertilizer is mixed upto the completion of seed surface coating. Sowing is done immediately after inoculation.
2.	Slurry method	Slurry is prepared by mixing the powder of biofertilizer into the solution of an adhesive, i.e. sugar, molases, or water soluble gum. It is mixed with seeds which are dried in shade. Sowing is done 8-10 hrs. later.
3.	Seed pelleting	Some pelleting material like powdered limestone, rock phosphate or basic slage is employed. Adhesive solution is than thoroughly mixed with seeds.
4.	Seed coating	Some coating material, i.e. charcoal, calcium silicate, diatomaceous earth, dyes of agro-chemicals is used along with some materials which stop flaking of coating (poly- vinyl alcohol, poly-urea varnishes, poly-urethane or resins dissolved in chemical solvents).

In tissue culture technique, a specific source is needed as explant, i.e., stem, hypocotyl, embryo, root segment, seeds etc. A drawback of this method is that it is very costly and requires a very high level of skill. Similarly, plant-lets prepared in the standardised environmental conditions of the tissue culture laboratory, when transferred to the field, they frequently die. But, the growth, in successful plants is always very fast and thus large scale afforestation and reforestation programmes are being launched throughout the country to restore the lost ecology and greenery by this method (Kandya, 1987).

Policies and Program

The National Wastelands Development Board of India has formulated some policies and programmes for the state forest departments and NGOs, through which a large area of wastelands is being converted into useful culturable and productive land. Social forestry, urban forestry, energy plantations, rural Forestly are such programmes which involve the masses in villages, Forest department employs persons to look-after the newly planted seedlings in the nearby areas, protect them from grazing and trampling, irrigate them and in return, they will get the products of that tree (fuel, timber or gum etc.) after a certain age of the plant. Not only this, the poor tribals and villagers are also paid some money every month for this job. Such an involvement of people is very effective But, unfortunately the rate of reclamation is still only 1/8 of the area becoming unproductive every year. This will need many more serious efforts if the welfare of the country is aimed.

Table 14. Success in the tissue culture grown plants achieved in India.

Species	Source of explant	Type of culture	Morphogenesis/ propagation
Albizia falcataria	Hypocotyl, stem	Callus	Shoot and/or root
Azadirachta indica	Embryos	Callus	Shoot buds, plant-lets
Dalbergia latifolia	Shoot/root segment		Direct Shoot formation
Eucalyptus tereticornis	Terminal buds	Shoot-tips	Multiple shoots
E. camaldulensis	Seedling nodes	Shoot-tips	Shoots & Roots
Leucaena leucocephala	Various parts	Callus	Plantlets
Paulownia tomentosa	Various nodes	Direct Buds & Planlets	
Populus ciliata	Fertilized ovule	Rooted Shoot tips Root Shoot proliferations	
Salix sps.	Lateral buds	Direct Shoot formation	
Stantalum album	Seeds, Embryos	Single Node rooted plantlets	
Tectona grandis	Bud meristem	Direct Root initiation	
Pinus radiata	Hypocotyl	Rot initiation	
Thuja plicata	Cotyledons	Multiple Budes with leaves	

REFERENCES

Anon, 1984. The state of India's environment. The second citizens report. Centre for Science and Environment, New Delhi.

Jadhav, R.N., A.K. Kandya and M.N. Tandon 1989. Mapping of potential afforestation sites through visual analysis of IRS-1A LISS-II data. *J. Trop. For* 5(3+4) : 67-71.

Kandya, A.K. 1987. Significance of seed technology in relation to the natural growth and development of forests and in reforestation programmes. Proc. MAB Regional Training Workshop. Colombo, Sri Lanka. p. 299-320.

Kulkarni, H.D. 1987. Vegetative multiplication of forestry trees by Mist Propagation Technique. Vaniki Sandesh, 11(1+2): 28- 30.

Prasad, R. 1988. Technology of Wastelands Development. Associated Publishing Co. Karol Bagh, New Delhi, p. 356.

Prasad, R. and A.K. Kandya 1989. Depletion of soil and forest cover and technology for control. Proc. National Seminar. SFRI Jabalpur, p. 277.

Prasad, R. and A.K. Kandya 1991. Handling of forestry seeds in India. Associated Publishing Co. New Delhi. p. 450.

Prasad, R. and P. Bhatnagar 1988. Grazing in forests : an ecological liability. SFRI Report no. 25, Jabalpur.

CHAPTER - 10

ENVIRONMENTAL PROTECTION THROUGH WASTELAND MANAGEMENT - A CASE STUDY OF COASTAL ORISSA

Malaya K. Misra

Ecology and Floristics Laboratory
P.G. Department of Botany, Berhampur University
Berhampur - 760 007, Orissa

INTRODUCTION

Land along with soil is a precious natural resource. Land provides majority of human needs such as food, fuel, shelter, fibre and medicine for its sustenance, but due to unprecedented increase in human population the pressure on land is gradually increasing. To meet the demand for food, the land is being used irrationally resulting then transformation into wastelands. This is a clear indication of mismanagement of land resources which not only reduces the productivity potentiality of the land but also changes the environment adversely.

The present communication deals with various categories of wastelands available in Orissa state, causes of formation and measures taken for the management of wastelands, particularly in the coastal areas.

DEFINITION AND WASTELANDS AREA

The wastelands are defined by different organisations in various ways in the Technical Task Group constituted by the

Planning Commission and National Wastelands Development Board (NWDB) define the wastelands as "the degraded and unutilised lands except the current fallows due to different constraints (CSIR, 1990)", which are further classified as culturable and nonculturable wastelands.

The official record of wastelands in India is about 175 million ha which is more than the half of the geographical area (329 million ha). However, the National Remote Sensing Agency in 1983 estimated the wastelands through aerial mapping to be about 55.41 million ha of which Orissa contributed about 1.95 million ha (CSIR 1990).

Land degradation is generally caused due to several environmental factors such as runoff, wind, deforestation, water logging, acidity and alkalinity. Various categories of land degradation. Maximum wastelands are created due to clearing of natural forests. The state is having much barren rocky as well as undulating uplands with or without shrubs.

Orissa coastline covers 480 km along the Bay of Bengal touching Balaswar, Cuttack, Puri and Ganjam districts (undivided). In Orissa coast, sand dunes poses much problem. The four coastal districts have about 55253 ha of sandy area, out of which 32% is lying bare, 52% is covered with thin vegetation (Table-1).

Table 1. Distribution of sand dunes in coastal Orissa.

Districts	Bare	Sand Dune Area (ha)		Total
		With sparse vegetation	With dense vegetation	
Balasore	1731	731	606	3068 (5.5%)
Cuttack	3784	1631	656	6071 (11%)
Puri	7502	24644	4181	36327 (65.2%)
Ganjam	4526	4431	831	9787 (17.7%)
Total	17542 (32%)	31437 (57%)	6274 (11%)	55253 (100%)

Figures in parentheses show percentage
Source : Mahapatra *et al.* (1984) URSAC, Bhubaneswar.

Sand dunes cover 0.35% of the total geographical area (155707 sq. km) of the state. Out of the four districts Puri district is having maximum sandy area. Flow of sand, from the coast, due to heavy wind casts sand on the interior agricultural lands. The depth of sand decreases with distance from sea (Nisanka 1991).

Further, 1.63% of the total area of the state, situated in the four coastal districts, is suffering from salinity (Table-2). Total saline land of Orissa is 254101 ha of which 36524 ha (14.4%) is lying idle without any use.

Table 2. Distribution of coastal saline lands in Orissa.
(Area in ha)

District	Land use			
	Agriculture	Mangrove	Idle	Total
Ganjam	6,480	—	3,465	9,925 (3.9)
Puri	43,930	3,250	13,076	60,256 (23.7)
Cuttack	67,342	18,667	9,440	95,449 (37.6)
Balasore.	75,679	3,249	10,543	88,471 (34.8)
Total	1,93,441 (76.1)	24,116 (9.5)	36,524 (14.4)	2,54,101 (100)

Figures in parentheses shows percentages.

As per the estimate of the conservator of forests, working circle, Orissa about 12000 sq. km of forests are either highly degraded or are completely barren and another 12000 sq. km of forests may be available for rehabilitation (Das 1992).

Shifting cultivation is also practised in the forests of the state by the tribals. The total area under this harmful practice as estimated by the Orissa Remote Sensing Application Centre (ORSAC) is about 184018 ha during 1984 (Table-3).

Table 3. Area under shifting cultivation in different districts of Orissa as assessed by Remote Sensing Agency (1984).

Sl.No.	Name of the District	Area (ha)
1.	Bolangir	1795
2.	Dhenkanal	5121
3.	Ganjam	21680
4.	Kalahandi	5862
5.	Keonjhar	40638
6.	Mayurbhanj	7303
7.	Koraput	30651
8.	Phulbani	20996
9.	Sambalpur	41781
10.	Sundargarh	8191
	Total	1,84,018

WASTELAND MANAGEMENT IN COASTAL ORISSA

Land management is a process to put the land to proper use. The major factors that influence land management are precipitation, percentage and length of slope, type of soil, nature of ground cover, erosion rate, wind velocity, rate of salinity or acidity or alkalinity and nature of ground cover. The classification of the land is generally done taking into account the stock of the situation of the land.

Plant cover plays a great role in reclamation of wastelands. For this keguminous plants, energy crops, economic important plants and other hardy species are generally selected. Species are selected on the basis of their acclamitazation in the existing, condition of soil. Casuarina and Cashew are the best suited species for coastal sands specially where nutrients status is very poor. The State Soil Conservation Department, Forest Department, Forest Corporation and some private agencies have been planting these two species on the coastal Orissa. The Botany Deptt. Berhampur University has also rehabilitate the costal wasteland by these two species and the details of strategies are discussed in the present text.

Casuarina (*C. equisetifolia*)

Most of the *casuarina* plantations in Orissa are in pure culture and exhibits good growth performance. It is also planted as guard species in the sandy coast to save the young cashew plants from wind and salt sprays. The area covered by Casuarine is 10112 ha of which the coastal districts cover 9451 ha (Table 4). This figure is an under estimate as plantations raised by private parties are not taken into account.

Table 4. Area under Casuarina and Cashew plantation in Orissa (upto 1987).

Sl.No.	Districts	Plantation Area (ha)	
		Casuarine	Cashew*
1.	Balasore	929	941
2.	Bolangir	—	681
3.	Cuttack	528	10796
4.	Dhenkanal	—	11366
5.	Ganjam	623	10168
6.	Kalahandi	—	1199
7.	Keonjhar	—	1643
8.	Koraput	—	10096
9.	Mayurbhanj	661	2230
10.	Puri	7371	16564
11.	Phulbani	—	623
12.	Sambalpur	—	6467
13.	Sundargarh	—	6305
	Total	10112	79079

* Excludes mixed plantation and private sector.
Source : Satapathy (1991).

Density of casuarina plants on sands varies greatly and in Ganjam coast it ranged from 2500 to 5000 trees per ha (Nisanka 1991 & Nisanka *et al.*, 1993) while in Puri coast it varied from 739 to 1608 trees per ha (Verma *et al.*, 1987). The relative growth rates based on diameter, plant height and dry weight, and leaf area ratio showed good performance. The total biomass production in these stands varied from 12.65 to 248.95 t ha^{-1} in 4 to 13 years in Ganjam coast (Nisanka, 1991) and 18.5 to 135.3 t ha^{-1} in 5 to 15 years in Puri coast (Verma *et al.,* 1987).

Litter production and nitrogen fixation are the two important properties of casuarina which help for the reclamation of the unfertile soil. Litterfall in *Casuarina equisetifolia* plantation continues throughout the year. Litter production varies between 8.7 to 29.6 t $ha^{-1}yr^{-1}$ in the plantations. Decomposition and nutrient studies revealed that nutrient standing states increased with age from 78.45 to 1052.59 kg ha^{-1} of nitrogen, 1.22 to 31.82 kg ha^{-1} of phosphorus, 28.47 to 536.07 kg ha^{-1} of sodium, 18.03 to 356.65 kg ha^{-1} of potassium and 122.35 to 1365.51 kg ha^{-1} of calcium. Nutrient release per hectare of plantation through annual litterfall was 313.7 kg N_2, 7.72 kg P, 74.05 kg Na, 59.24 kg K and 553.89 kg Ca (Misra & Nisanka 1997). Although *C. equisetifolia* is a non-leguminous plant it fixes the free atmospheric nitrogen through *Frankia* (an actinomycetes) in root noddules. It has been reported that this species fixes a minimum of 58 kg N_2 ha^{-1} $year^{-1}$ (Dommergues 1963). Due to this property the species enriches the soil with nitrogen. The improvement of soil was also observed through the colonisation of herbaceous species in ground vegetative cover and productivity of earthworms under casuarina and coconut plantations in coastal Orissa, which revealed the improvement of soil quality and the environment.

Thus *C. equisetifolia* contributes a lot in changing the micro-climate of the area and improving the soil quality by increasing the water-holding capacity, organic matter and organic carbon of soil (Nisanka 1991). It serves as wind break and checks interior flow of sand and to some extent cyclone.

Cashew

Cashew (*Anacardium occidentale*) an exotic species (native of Brazil) has been successfully raised in the coastal Orissa, on sands and hillocks, which helped in improving wastelands. Till 1987-88, the cashew plantations raised in different districts by the state government was 79079 ha (Table 4). However, this value goes upto 150343 ha when mixed plantations and plantations raised by private sector are taken into accout. Cashew plants due to its soil binding capacity, helps enormously in conserving the soil. Plantations in catchment areas helped reduce the inflow of silt into the major and minor reservoirs and reduced the intensity of flood in the state. In coastal area it checks the wind erosion and act as a wind break against cyclone and tornado. The low spreading habit and criss-cross branching pattern help the cashew plant to arrest the sand drifting and fixes sand dunes in its original position (Patro & Behera 1979). Cashew plantation in Orissa offers roughly 1135 million square metre of green canopy cover to the bald barren lands (Satapathy 1991). This increases evapo-transpiration and thus decrease the ambient temperature of the area.

Cashew plantation enriches the soil through litter production and decomposition and creats a conducive habitat for the growth of soil macro- and micro-organisms (Behera, 1989). This enhances the fertility level of the soil. Due to all these qualities, cashew plantations in the state, gradually converts the wastelands into productive cultivable lands.

Screwpine (*Pandanus fascicularis*)

It grows naturally on the coastal wastelands it is vegetatively propogated by the local inhabitants for soil conservation, as hedge plant and above all for its fragrant flowers. Presently a single flower costs about Rs. 4 to 6. This species is abundantly present on sands, field margins, banks of estuaries and rivers, and road sides in coastal areas. Screwpine is a good sand binder and a salt resistant species.

Coconut (*Cocos nucifera*)

Plantations are largely observed in the coastal districts of Orissa. It well adopts to the coastal climate and improves the environment.

REFERENCES

Behera, N. 1989. *Annual Project Report* on "Micro-flora, litter decomposition and soil fertility status in Cashew plantations of Orissa under varied agro-climatic conditions. Ministry of Environment and Forests, New Delhi.

CSIR, 1990. *Plants For Reclamation of Wastelands*. Publication and Information Directorate, Council of Scientific and Industrial Research, New Delhi.

Das, S.S. 1992. The status of forests in Orissa. pp. 10-28. *In : Debate and Issues in Social Forestry in Orissa*. CENDERT, SIDA and ISO-SWEDFOREST, Bhubaneswar.

Dommergues, Y.R. 1963. Evaluation due taux defixation de l'azotedans un sol dunaire reboise en filau (*Casuarine equisetifolia*). *Agrochem.* 7 : 335-340.

Misra, M.K. and M. Ray. 1995. Phytosociology and Earthworm production in the groundfloor of Casuarina and coconut plantations in coastal Orissa. pp. 165-176. In : P.C. Misra,

N. Behera, B.K. Senapati and B.C. Guru (eds. *Advances in Ecology and Environmental Sciences*. Ashish Publishing House, New Delhi.

Misra, M.K. and S.K. Nisanka, 1997. Litterfall, decomposition and nutrient release in *Casuarina equisetifolia* plantations on the sandy coast of Orissa, India. *Tropical Ecology* 38(1) : 109-119.

Nisanka, S.K., S.P. Nayak and M.K. Misra 1993. Primary productivity and growth performance of *Casuarina equisetifolia* : pp. 89-93. In : N.P. Melkania (ed.) *Social Forestry and Wastelands Development in Mid-Eastern India*. Indian Institute of Forest Management, Bhopal.

Patro, C. and R.N. Behera 1979. Cashew helps to fix sand dunes in Orissa. *Indian Farming* 28(12) : 31.

Satapathy, H.K. 1991. Ecological and Economic aspects of Cashew plantation in Orissa. *Ph.D. thesis*, Berhampur University, India.

Verma, V.P.S., V.N. Tandon and H.S. Rawat 1987. Biomass production and plant nutrient distribution in different aged plantations of *Casuarina equisetifolia* in Puri, Orissa. *Indian Forester* 113 : 273-280.

CHAPTER - 11

REVEGETATION OVER PYRITE WASTE DUMP OF AMJHORE MINING AREAS

Kumar Birendra* **& R.S. Bilgrami****

*Scientist, CMRI, Dhanbad-826001

** Deptt of Botany, BRA Bihar University, Muzaffarpur-842 001

INTRODUCTION

Pyrite content in waste dumps of coal mines in India is generally low except those in Assam and Meghalaya (total sulphur 8%), Wardha Valley (total S 2-4%), Nowrazabad colliery SECL, (pyritic S/ 1-3%, sulphate S 20-77% and organic S 5-15%). Apart from this, the coal mines facing AMD problems are at the Yellandu area of Singareni coalfield in Andhra Pradesh. The semianthracite coals of Jammu and Kashmir is also reported to contain 9% sulphur with organic sulphur of 1% or less. Mining for base metals has increased manifold in India during the last decade. At Zawar, Rajpura Dariba, Rampur Agucha and Sargapalli lead-zinc mines (HZL) heaps of tailings, having sulphur content upto 22%. However, the most serious AMD problem in India is being faced at Amjhore pyrite mine, located in the district of Rohtas (Bihar), which is one of the most important sources of pyrite in the country. During the course of mining, top and bottom shales are also excavated, diluting the grade of ROM from the 40% to 22% sulphur. The low grade sedimentary pyrite contains 10-12% sulphur and is highly liable to spontaneous heating. By coming in contact with moisture and rainwater, it tends to produce weak sulphuric acid. Shales, produced in the course of mining, were being manually picked and dumped at different places in around 4,000 m^2 area. These old dumps roughly contain around 0.3 million tonnes of material. They had been undergoing biochemical oxidation, cought fire and have got burnt. These burnt out shales still contain high percentage of sulphur. During the rainy season, water percolates through these shale dumps. Several samples of water collected by the authors indicate their high acidity and higher concentration of metallic ions.

Only a few probing studies have been conducted in India on abatement of acid mine drainage (AMD) and environmental management of such acidic waste dumps. Authors have conducted comprehensive studies on the interaction of the acidic environment with plant growth by buffering its acidity with phosphatic waste in the ratio of 3 : 1 under laboratory trials. At this ratio the prolific growth of plants was observed (Thakur *et al.*, 1992). Keeping this in view investigations were extended to the field trials. This paper relates to the field study conducted at one the dumps at Amjhore pyrite mine.

EXPERIMENTAL PROCEDURE & FIELD STUDY

The chemical analyses of the old burnt out pyritic waste and phosphatic waste was done in CMRI laboratory following the standard procedure using Atomic Absorption Spectrophotometer (Model 902, GBC). The results are given in Table-1 which shows that pyritic waste is highly acidic and contains higher concentration of heavy metals.

Table 1. Chemical composition of pyrite waste and phosphatic waste.

Chemical Constituents	Pyrite Waste	Phosphatic Waste
pH	3.30	8.09
Fe	70,310.00	26,005.00
Mn	307.36	7.30
Cu	128.90	26.90
Zn	223.50	3.90
Pb	83.40	34.50
Ni	125.70	10.20
Co	32.40	3.00
Cd	2.00	0.00

Concentrations are in g/g except pH.

Dump preparation and reclamation

A reclamation strategy was worked out for preventing ingress of oxygen and rainwater with a view to controlling further oxidation and weathering of the pyritic waste material and soil erosion from this dump. For finding the effectiveness of the reclamation procedure one such dump (about 200 m^2) was selected for experimentation (Fig. 1). This experimental dump was reshaped and terraced to a slope of 1 in 3 after laying a thin layer of phosphatic waste. 75 mm layer of top soil was also provided as the topmost layer of the benches. Water drainage arrangement was made on the side of the benches along the slope of the terraced dump down right upto the roe so as to drain out the rainwater without seepage into the dump.

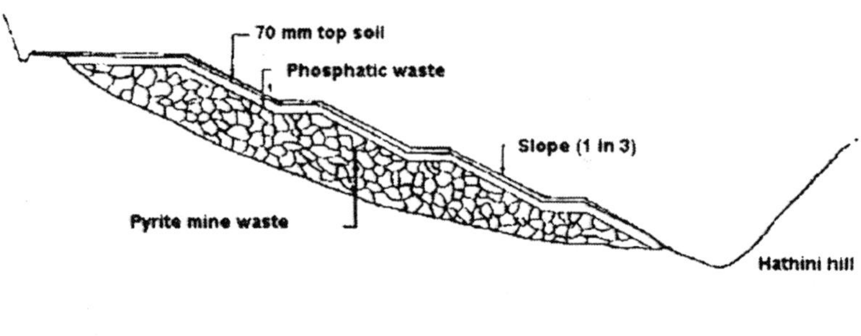

Fig. 1. Scheme of Terracing and Drainage Arrangement over Pyrite Mine Waste

Biomass generated over ameliorated pyritic waste

Pure seeds of *Crotalaria juncea* (Sunnhemp family Leguminoceae), *Pennisetum typhoides* (Bajra family Poaceae), and *Ricinus communis* (Castor family euphobiaceae) were procured from the standard seed store. Before sowing they were disinfected with 0.001% mercuric chloride solution for one minute and thereafter thoroughly rinsed with water. They were seeded in the soil digging holes of 2.5 cm depth. Seeds were sown equidistantly over the ameliorated terraces leaving one bench for individual species to monitor the growth of the plants and tolerance capacity. The Lowermost bench was seeded with mixed flora to monitor the combined growth potential. Since experiments were conducted in the monsoon season, no arrangement for sprinkling of water was made. However, during prolonged dry periods, occasional water sprinkling with the help of water tanker was done.

For monitoring the biomass, plants of different species were harvested from specified areas after 45 and 78 days, dried at 60°C, weighed till constant weight was achieved.

The biomass were measured in all the above mentioned plant species (Fig. 2 & 3). Heavy metal constituents in the plant material were determined (oven dried at 80°C for 48 hours, ground and digested in acid by Atomic Absorption Spectrophotometer. The results of heavy metal translocated to the different parts of the plants have been shown in Table-2.

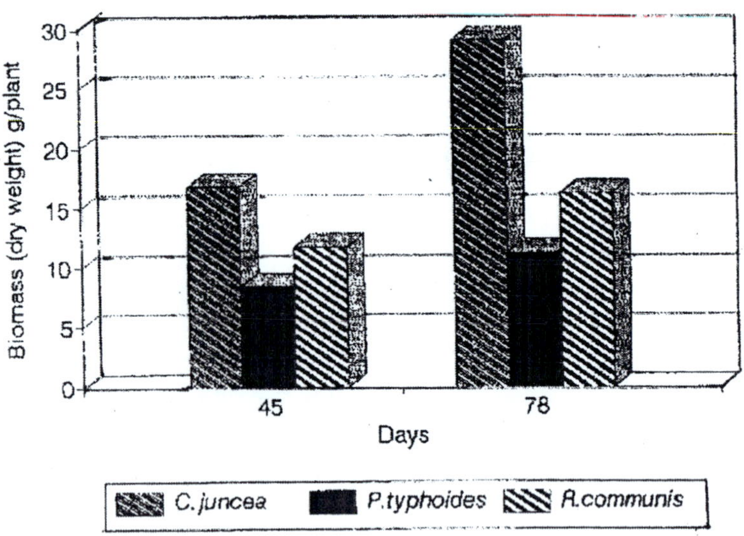

Fig. 2. Root biomass generated over amended pyrite mine waste dump

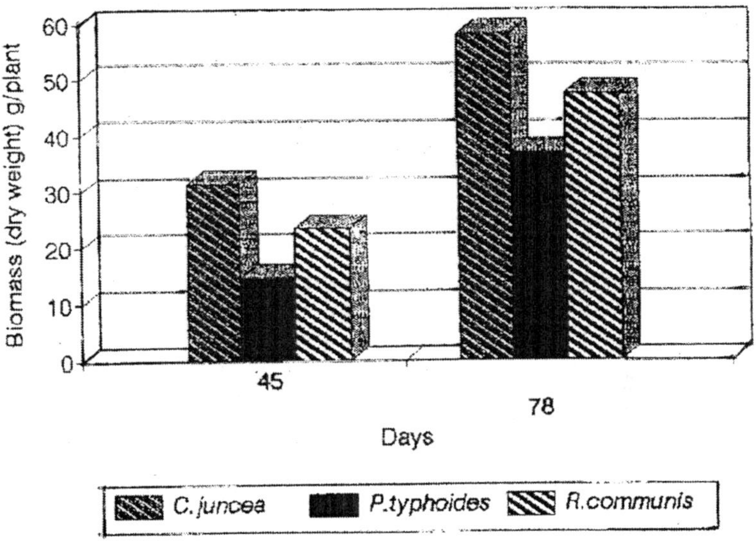

Fig. 3 : Biomass generated over amended pyrite mine waste dump.

Table 2. Concentration of metals in different parts of the plant (g/g dry weight)

Name of the Plants	Parts of the Plants	Cu	Zn	Fe	Mn	Co	Pb	Cr
Ricinus communis	Root	12.27	37.30	332.12	25.49	1.95	5.50	4.29
	Shoot	147.90	31.47	199.03	25.85	1.67	3.44	2.00
	Leaf	258.90	63.13	69707.23	1643.00	8.96	45.80	37.84
Crotalaria juncea	Root	24.79	90.01	1385.86	45.51	17.32	27.17	61.14
	Shoot	16.05	20.67	450.30	14.38	0.64	4.17	1.96
	Leaf	622.09	210.17	55668.60	1686.04	0.03	158.43	263.00
Pennisetum typhoids	Root	17.73	79.87	7978.00	188.00	5.55	0.35	50.00
	Shoot	8.82	48.50	223.43	53.83	20.81	2.18	0.82
	Leaf	35.62	73.94	757.28	51.94	22.94	3.24	5.20

Translocation of heavy metals in plants

The total metal content of the soil and/or air is not fully available to a plant. The entry of metals into plants depend upon their extractability potential or the ease of exchange of the root soil interface. So, each metal will have a different degree of availability depending upon various factors, like the nature of the metal compound, condition of the soil such pH, presence of competing ions, chelating agents, and also micro-organism. Organic matter can form stable complexes with heavy metal ions, rendering them unavailable for uptake by the plants (Hudson *et al.*, 1966, Dykeman and de Sousa, 1966). Other ions also affect the availability or uptake of heavy metal ions. Ion concentrations are varied and complex and our understanding of these interactions is still not complete (Foy *et al.*, 1978).

Our knowledge of the heavy metal translocations to and accumulation in plant tissues is based mainly on studies involving tissue analysis such as leaves, twigs and bark for the metal. Heavy metals appear to be transported as ionic complexes (Bowen, 1979). In some cases specific organic or amino acid chelaters may be involved (Foy *et al.*, 1978). Interaction among plant nutrients within the symplast affect translocation.

RESULTS AND CONCLUSION

After 78 days of experimentation, the soil pH level had remained 6.34 against 3.30 of the normal unameliorated pyritic waste. Data obtained from plant species grown an amended mixture and top soil significantly increase plant biomass. After 45 and 78 days of harvesting *C. juncea* showed greater aboveground biomass (31.43 and 58.19 g/plant) in comparison to *P. typhoides* (14.43 and 36.73 g/plant) and *R. communis* (23.32 and 47.29 g/plant). The biomass of the plant species showed the trend *C. juncea, R. communis, P. typhoides* (Fig. 2). The highest rate of root biomass (16.83 and 29.12 g/plant) was noticed in *C. juncea* after 45 and 78 days of interval (Fig. 3). Whereas, in *R. communis* was 11.76 and 16.32 g/plant. Root biomass was lowest in the plant *P. typhoides* (8.41 and 11.32 g/plant). In the light of above parameters *C. juncea* plant exhibited more prolific growth than other two plant species.

Translocation of heavy metals in plants have been studied by several authors (Salisbury and Ross, 1978; Bowen, 1979; Foy *et al.*, 1978 & Antonovics *et al.*, 1971). In their view most heavy metals are not highly mobile than lead (Antonovics *et al.*, 1971). When copper and zinc are present in the soil at normal concentrations, the actively growing parts, leaves and branches have the highest copper and zinc concentration, roots and barks have intermediate concentration and wood has the lowest concentration. Lead concentrations are almost always highest in roots (Malyuga, 1964; Ernst, 1972; Jones *et al.*, 1973). When heavy metals are present in excessive concentrations, the highest tissue concentration tend to be highest in woody stem, and zinc is likely to be highest in leaves or flower (Lane and Martin, 1977; Larcher, 1975).

In the light of above observations, when results are compared, it could be observed from Table-2 that these three plant species do not always show the highest concentration of the heavy metals in the root, rather the actively growing parts of the plant specially leaves, show the highest concentration of the heavy metal.

Concentrationwise, iron is the most translocable metal in all parts of the plants. Its concentration is the highest in the leaves of *R. communis* and *C. juncea* going upto 69707 g/g in the leaves of *R. communis*. However, the roots of *P. typhoides* concentrate the highest amount of iron compared to its growing parts, like the shoot and leaves. This is due to the high concentration of this material in the phosphatic waste as well as in the pyritic waste Table-1.

The next most translocable metal was found to be manganese. As in case of iron, its highest concentration was found in the leaves of *R. communis* and *C. juncea* with roots of *P. typhoides* concentrating its highest concentration. Pyrite waste of Amjhore (Table-1) was the highest source of supply of manganese to these plants.

The next translocable metal is copper, which got concentrated in the leaves of all the three plants with leaves of *C. juncea* showing the highest concentration of copper (622 g/g) compared with only 35.6 g/g in case of *P. typhoides*. The root and shoot of these plants show intermediate to low concentration.

Compared to copper, zinc appears to be less translocable and its highest concentration was found in the leaves of *C. juncea* (210 g/g). Zinc can function as a co-factor with several roles; it may be necessary for enzyme activity, enzyme regulation or any combination of these functions (Clarkson and Hanson, 1980). Zinc is not involved in redox reaction, but can in the presence of NAD, catalyse dehydrogenase reactions (Bowen 1979). Zinc seems to be essential for the stability of cytoplasmic ribosome (Bowen, 1960; Clarkson and Hanson, 1980).

Lead is not an essential element for plant growth, but is taken up by plants from soil as well as from atmosphere. It may enter passively or may be accumulated via specific uptake mechanism if it resembles certain essential element in its physical or chemical properties (Salisbury & Ross, 1978). At very low (normal) concentration, lead appears to have effect on plant growth. It often unavailable for the uptake. Our studies indicate that its concentration is available in the leaves of *R. communis*, *C. juncea* and *P. typhoides* with root and shoot showing very low to intermediate concentration. In fact, lead concentration in the leaves of *C. juncea* was as high as 158 g/g, but in case of *P. typhoides* it was only upto the order of 3 g/g.

Cobalt concentration was also the highest in the leaves of *R. communis* and *P. typhoides* but the roots of *C. juncea* also contained appreciable concentration of cobalt (17 g/g). From the above analysis, it appears that the leaves of the three plant species, growing on soil containing high concentration of heavy metals and the above ground parts contain very high concentration of heavy metals. It is, therefore, imperative that the reclaimed area must be effectively fenced so as to prevent grazing by cattle. The present investigation does not necessarily reflect the actual physical load of the contaminants but provides a key to understand its impact on the ecosystem.

REFERENCES

Salisbury, F.B. and C.W. Ross 1978. Plant Physiology Second edition. Wodsworth Publishing Company, Incorporated, Belmont, California.

Gilbert, F.A. Copper in nutrition, 1952. Advances in Agronomy 4 : 147-177.

Antonovics, J.A.D. Bradshaw & R.G. Turner 1971. Heavy metal tolerance in plants. Advance in Ecological Research 7 : 1-85, 1971.

Behra, B.N., B.B. Panda 1989. Plants to monitor metal pollution, Science Reporter, April-May.

Hudson, J.F., W.F. Lindsay and J.F. Frierweiler 1966. Micronutrient complexing in soil solution II. Complexing of zinc and copper in displaced solution from calcarious. Proceedings of the soil science society of America 30 : 723-726.

Dykeman, W.R. and A.S. de Sousa 1966. Natural mechanisms of copper tolerance in a copper swamp forest. Canadian Journal of Botany, 44 : 871-878.

Foy, C.D., R.L. Chaney and M.C. White 1978. The physiology of metal toxicity in plants. Annual Review of Plant Physiology. 29 : 511-566.

Bowen, H.T.M. Environmental Chemistry of the Elements 1979. Academic Press, New York.

Clarkson, D.T. & J.B. Hanson, 1980. The mineral nutrition of higher plants. Annual Review of Plant Physiology 31 : 239-298.

Ernst, W. 1972. Eccophysiological studies on heavy metal plants in South Central Africa, Kirkia 8 : 125-145.

Jones, L.H.P., C.R. Clement, M.J. Hopper 1973. Lead uptake from solution by perennial ryegrass and its transport from roots to shoots plant and soil, 38 : 403-414.

Lane, S.D. and E.S. Martin 1977. A histochemical investigation of lead uptake in Raphanus sativus. The New Phytologist 79 : 281- 286.

Larcher, W. 1975. Physiological plant ecology Springer-Verlag, Berlin Translated from the Germal by M.A. Biederman-Thomson.

Malyuga, D.P. 1964. Biogeochemical methods of prospecting, Consultants Bureau, New York.

CHAPTER - 12

MINING AND WASTELAND DEVELOPMENT

S.K. Singh
University Department of Chemistry
T.M. Bhagalpur University, Bhagalpur

INTRODUCTION

During the last three decades, the world industrial output has increased more than six times while the world population has merely doubled. The scale and level of requirements of minerals have greatly increased and we are heading towards the stage where a much larger consumption of minerables will be inevitable. Extraction and development of minerals are closely linked with other national resources such as land, water, air and forests. Mining activity often leads to environmental problems like land degradation and wasteland generation, deforestation, atomspheric pollution, disposal of solids and wastes leading to acid-mine drainage etc. affecting the ecological balance of the surroundings.

MINING PROJECTS AND ENVIRONMENT

A mining project involves the following stages

- Preliminary and detailed exploration of deposit and evaluation of the exploration report.
- Baseline measurement of environment parameters and preparation of environmental impact assessment.

- Preparation of preliminary and detailed techno-economic feasibility report followed by the detailed project report based on above.
- Development of infrastructure and preliminary site development and setting up of site facilities.
- Erection of plant and equipments.
- Preproduction development of mine followed by regular production.

The preliminary and detailed exploration have a minimum impact on the environment. Only the construction of approach reads the preparation of the drilling sites require deforestation in some cases leading to soil erosion. Land degradation is the most serious environmental impact of open cast mining. Real problem surfaces after mine starts regular production. The toxicity of the soil is caused by oxidation of pyrite which produces salts and makes it highly acidic.

THE GENERATION OF ACID FROM PYRITIC MATERIALS

The oxidation of pyritic materials associated with coal and other mined minerals occurs according to the following reactions :

1. $FeS_2(S) + 7/2O_2 + H_2O \longrightarrow Fe^{2+} + 2SO_4^{2-} + 2H^+$
2. $Fe^{2+} + 1/4O_2 + H^+ \longrightarrow Fe^{3+} + 1/2H_2O$
3. $Fe^{3+} + 3H_2O \longrightarrow Fe(OH)_3\ (S) + 3H^+$

The overall stoichiometric equation is

4. $4FeS_2(S) + 15O_2 + 14H_2O \longrightarrow 4Fe(OH)_3 + 8H_2SO_4$

Calculations show that for material containing 0.1% sulphur, all as pyrite, complete oxidation will yield a quantity of sulphuric acid that will require 6,250 ponds of calcium carbonate to neutralise 1000 tons of materials. This is the same as 0.0624 meq of H^+ per gm of material or if one gm of this material and the acid derived therefrom were placed in one ml of water, one would obtain a solution of 0.0624 N H^+ and a pH of 1.20, theoretically.

Thus, if only one tenth of the pyrite in the above example is oxidised, a resultant solution will have a pH slightly above 2. Thus the quantity of acid, corresponding to low pH solutions, which results from only a small amount of pyrite oxidation can be seen to greatly influence the acidity of a moist spoil area.

Equation (1) represents the oxidation of crystalline pyrite by oxygen, which is the critical reaction. The reaction sites are the pyrite crystal surfaces exposed to oxygen, equation (2) and (3) represent the oxidation of ferrous iron to ferric iron and the subsequent precipitation of ferric iron as ferric hydroxide, giving the familiar red colour observed in acid streams. These reactions do not in general occur at the pyrite surfaces, and thus have little or no bearing on the rate of pyrite oxidation.

CONTROLLING FACTORS OF RATE OF PYRITE OXIDATION

For a given type and particle size of pyrite, the rate of reaction represented by equation (1) is a function of oxygen concentration, temperature, degree of surface saturation by water and pH of the solution in contact with the pyrite (Smith and Shumate, 1971). The conditions at the pyrite surface control the rate of the oxidation reaction.

Fig. 1 shows the effect of oxygen concentration on pyrite oxidation rate. The oxygen concentrations are oxygen dissolved in the water phase on the surface of pyrite. These data show that as the oxygen concentration is halved, the oxidation rate of exposed pyrite is reduced to approximately 60% of the rate of air oxidation. Oxidation rate becomes zero in absence of oxygen.

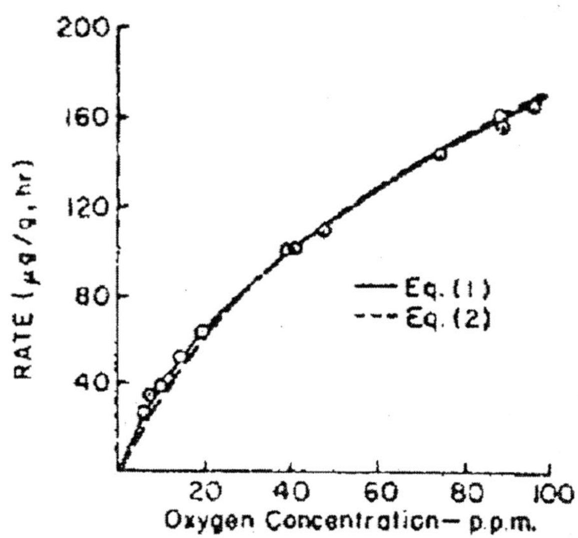

Fig. 1. Oxidation Rate vs Oxygen Concentration in Aqueous Phase

Fig. 2 shows that the rate of pyrite oxidation is related to the degree of saturation of the pyrite surface, reaching a maximum when the surface becomes completely wet. This does not mean complete immersion of pyrite in liquid water. The oxidation rate of pyrite exposed to air of 100% relative humidity is essentially the same as pyrite immersed in oxygen saturated water.

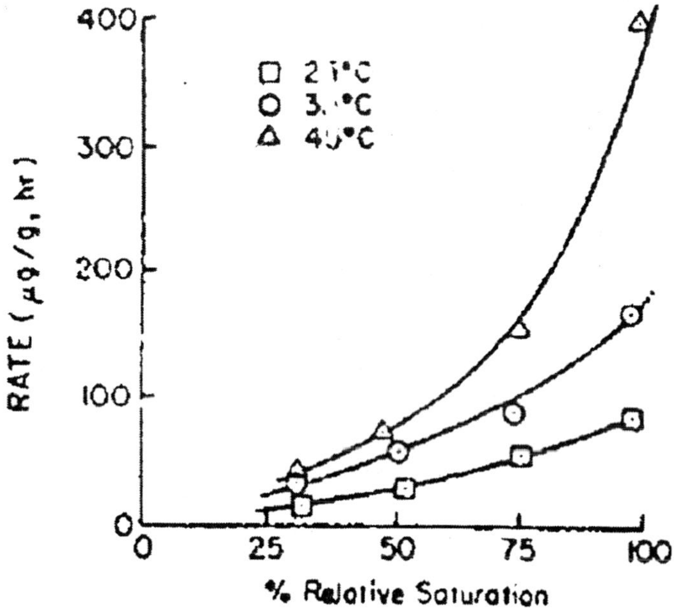

Fig. 2. Oxidation Rate vs % Relative Saturation

Fig. 2 also shows the effect of temperature on oxidation rate. For every 10°C rise in temperature, the rate approximately doubles- a characteristic of chemically controlled reactions.

Effect of pH of the solution in contact with pyrite on the rate of pyrite oxidation is shown in Fig. 3. The oxidation rate increases as pH increases and decreases as the solution becomes more acidic. The most important contribution this observation makes to acid mine drainage abatement is that liming or dusting the walls of a mine with lime will actually increase pyrite oxidation, although acid formed will be neutralised by lime.

In addition microbial catalysis of pyrite oxidation may contribute to acid mine drainage. Since essentially all reaction sites are exposed to a vapour phase, the amount of water in contact with the reaction sites is very small. This water is also a highly concentrated solution in both sulphate salts and sulphuric acid. Laboratory results (Smith and Shumate, 1971) show that microbial catalysis under these conditions is insignificant compared to the normal chemical oxidation rate. Bacteria do influence the discharge water by oxidising Fe^{2+} to Fe^{3+} in the ponded water within the mine, but they perhabs have no significant effect on the rate of formation of acid mine drainage.

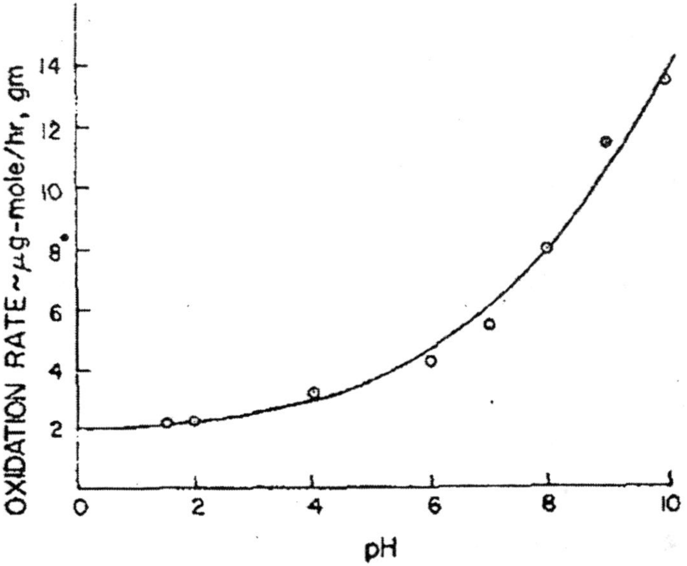

Fig. 3. Oxidation Rate vs pH

SPOIL COMPONENTS THAT MAY MODIFY THE EFFECT OF OXIDISED PYRITE

1. Basic constituents

Appropriate basic compounds present in spoil materials are (a) Salts of weak acids such as carbonates, (b) inherently basic compounds such as oxides and hydroxides. These basic compounds undergo neutralisation reaction with mine spoil acid resulting in an increase in exchangeable calcium.

Finer grades of lime-stone are much more rapidly effective for the purpose (Table 1 : From Proceedings of Acid mine drainage workshop, 1971, pp. 140).

Table 1. A comparison of various grades of linestone when applied at the same rates, Pennsylvania Agricultural Experiment Station (100 - mesh material is taken as 100)

Conditions	100 mesh and Smaller	60-80 mesh	20-40 mesh	8-12 mesh
Value in correcting acidity	100	57	27	18
Formation of nitrates	100	94	56	12
Plant Growth	100	69	22	05

The author also indicates that while coarser lime is less rapid in its action, it remains in the soil longer and its influence should be effective for a greater period of years.

II. Buffering Capacity

Clay minerals comprise a large proportion of the colloidal clay size particles in most soils. They have large surface area, are also very adhesive and have the property of swelling when dampened and of shrinking upon loss of water.

The properties of the minerals are influenced considerably by the quantity and nature of the cations adsorbed on them. Because of the high activity of clay minerals, a comparatively small quantity of them will influence greatly the physical and chemical properties of a soil. As long as the negative charges of the clay are largely satisfied by basic ions the soil exhibits little acidity, but as the basic ions are lost through leaching or plant use and are replaced by hydrogen ions the soil increases in acidity. There are two major factors causing large changes in soil pH :

(1) those which result in increased adsorbed hydrogen

(2) those which increase the content of adsorbed bases

In mine spoil materials, the source of hydrogen ions is the sulphuric acid formed by oxidation of pyrite. A large quantity of clay mineral colloids being present in spoil material will tend to absorb more H^+ ions from an acidic soil solution, and likewise release into the soil solution the bases associated with the colloids and thus resist somewhat a change to pH of the soil solution. Buffering capacity in spoil materials is related to the composition of the original parent rock material. So shales and clays can be expected to offer greatest resistance to pH change when acid generation commences, and in some cases may prevent high acidities from developing for a sufficiently long time to allow addition of lime needed to neutralise the acid, thus maintaining a soil environment favourable to vegetation from the beginning of exposure.

AT-SOURCE CONTROL OF PYRITE OXIDATION

In any mine drainage, there is a choice between at-source control of pollution and drainage treatment. Because pyrite oxidises rather slowly in nature, drainage treatment has to be carried out for decades after the mining operation has ceased. So it is advisable to apply at-source control wherever possible.

The key to the control of pyrite oxidation is the control of oxygen concentration at the pyrite surfaces. The oxygen exclusion principle can be effectively used to develop wasteland generated by mining activities. Refuse piles accumulated after mining contains pyrite. Sediment and acid drainage are the primary pollutants from such systems. The reacting zone of pyrite is limited to the top several inches of the pile, with the clay fractions acting as effective diffusion barriers against oxygen penetration to deeper levels. Between rains, the acid salts (oxidation products) build up in the outermantle. When it rains, much of the acid and iron is carried away in direct run-off, and some percolates down into the pile interior, to eventually seep out around the pile perimeter. The pH of the drainage fluctuates around 3.0. The oxidation goes on almost continuously. The most effective way to combat it is 'self healing' vegitative covers. The pile may be regraded, limed to neutralise surface acidity and then covered with several feet of sweet soil which is fertilised and seeded. The vegitative cover will prevent further exposure of pyrite, and hence will stop the oxidation reaction.

RECLAMATION OF SPOIL BANKS

Stacking of waste generated during mining causes considerable land degradation. Mining projects have adverse impact on flora and fauna and invariably cause deforestation Afforestation carried out by mining operators have given encouraging results.

In a study conducted by Sutton (1971) at Ohio Agricultural Research and Development Centre on the acid and toxic spoil materials, it was suggested that sufficient lime-stone be mixed so that pH of the spoil-bank increases above pH 6.0 and then the wasteland should be seeded with sweet clover, redtop, orchard grass etc. In another study, sweet soil was used to cover a highly acid spoil bank (pH 2,3). Plants like sweet clover, Korean Lespedeza grew in the areas covered with 4, 5, 8 and 10 inches of soil.

With spoilbanks that are acid but not toxic, selection of plant species will be helpful in establishing vegetation. Weeping lovegrass and redtop made the best growth in a sandy spoil material with a pH of 4, 3, of the legumes, sweet-clover made the best growth at pH 4, 3.

Spoilbanks in general are low in nitrogen. Some spoilbanks test low in either phosphorous or potassium or both elements. In an investigation using straw mulch at the rate of 2200 pounds per acre, a more dense plant cover was obtained on a sandy spoil. Mulch reduces evaporation of water from the surface thereby conserving moisture. By conserving moisture salt concentration will remain at a lower level in the spoil solution. Mulch will help maintain a more favourable moisture content at the surface which is important for a small seedling.

The dark spoilbanks have higher temperatures at the surface. These temperatures may become high enough to damage young seedlings.

Ghosh and Ghosh (1992) suggested regarding already subsided areas in Jharia Coalfield that these may be filled with domestic garbage, and covered with shale, clay or mulches made of grass, bush or weed. The total mass, after some years of composting, will form a fertile land most suitable for greening.

ENVIRONMENTAL POLICY AND MANAGEMENT

Being exhaustible and non-renewable, the exploitation of minerals needs to be carried out in a systematic and rational manner keeping in mind not only the short but the long-term perspective for the overall development of the country. The concept of control and command of the industry by the Government through environment regulations has now become an indicator of ultimate utility in advanced countries. Detailed legislative measures have been enacted to deal with the adverse effects of mining operations on environment and forests in India. A conscious strategy for the development of mineral resources constitutes the core of the minerals policy of the country pronounced in March, 1993 which clearly lays down the objectives of mineral development with due emphasis on their conservation and protection of the environment. The mining plan, inclusive of the environment management plan, prepared on the basis of guidelines issued by Indian Bureau of Mines (JBM) is mandatory for all mining projects. Consequently, there has been a remarkable improvement in the status of environment management in metalliferrous mines in India.

Sesa Goa Limited has been the topmost producer and exporter of iron ore in the Indian private sector. It has now diversified into different areas. It launched afforestation programme by planting local cashew in the sixties at its Orasso Dongor mine in the Advalpale village. Today, this mine has developed into a fully grown cashew orchard, giving a good yield to the locals. From 1982 onwards, the company was more systematic in the efforts and established its first nursery at the Sanquelim mine. Again in 1994, one more nursery was set up at codli mine (Sawant *et al.,* 1997).

To utilise the mine rejects for more productive purposes, the agri-horticultural approach was adopted on an experimental basis. Initially a thick plantation of Acacia species was grown which is a fast growing leguminous tree and not grazed by cattle. This vegetation cover protects the soil from erosion, helps in land stabilisation and the building up of organic matter content. Once the entire area stabilises (after about 4 yrs.), cash crops like cashew, jackfruit, coconut, banana, mango etc. are planted in between by cutting branches of the Acacia trees to the extent needed. The experiment has shown encouraging results.

In the last one decade, about two million plants (bio-diversified multispecies) have been planted in Tisco mines over an approximate area of 400 hectares.

The mining plan also promotes social forestry. With the introduction and implementation of the mining plan integrated with Environmental Management Plan, IBM has lived upto the expectation by way of making concerted efforts to improve the status of the mines in a holistic manner.

The Mathura refinery has shown that an effective environmental management system can substantially mitigate adverse impacts. It has the unique distinction of being the first refinery in Asia to be certified under British standard 7750 (ISO 14001) for its Environmental Management System. The vision of 'Green Refinery Clean Refinery' has come into focus through eco-friendly management programmes and practices adopted since its inception stage itself.

Environment is a key issue concerning the world as a whole. As issues have become increasingly interconnected and more intractable, it has become clear that global environment threats can be managed best by a more integrated and equitable approach to economic and social development.

REFERENCES

Ghosh, R. and D.N. Ghosh, 1992. 'Subsidence in Mining areas, its cause, impact and protection' the case of Jharia Coalfield'. Journal J. AEM, Vol. 19 No. 1, pp. 28-31.

Sawant, S.M.C.S. Dhaveje, M. Patil, 1997. 'Environmentally sound mining practices at Sesa Goa Limited'. G.B.O. Vol. 3, Issue 2, pp. 24-25.

Smith, E.E. and K.S. Shwmate, 1971. 'Rate of pyrite oxidation and acid production rate in the field'. In Proceedings of Acid Mine Drainage Workshop, pp. 11-18.

Sutton, P. 1971. 'Reclamation of Toxic Spoilbanks' In : Proceedings of Acid Mine Drainage Workshop, pp. 101-102.

CHAPTER - 13

FLORA OF WASTELANDS

P.K. Mukherjee

Department of Botany, Calcutta University
Calcutta - 700 019

INTRODUCTION

Any land which is not being used to its full potential is called a wasteland (Burt & Bradshaw, 1986). A wasteland can result from inherent or imposed disabilities such as by location, environment, chemical and physical properties of the soil or by financial and management constraints. Wastelands can be classified either as derelict, neglected and/or operational or as culturable and non-culdurable.

A neglected land is one which though capable of some beneficial use is at present uncared for, untidy and in a condition detrimental to the environment. That type results chiefly from transformation of forest land for some other use and its subsequent decline and disuse. A vast proportion of wastelands in any country belongs to this category.

A derelict land on the other hand is one which being so damaged by industrial and other development is incapable of beneficial use without treatment. The damage may be caused by both extinct or extant causes and thus be called discussed derelict land or operational derelict land (Wallwork, 1974). The total area of dereliction in any country is quite small, and in India it covers 0.04% of the total geographical area (NRSA, 1993). The crucial feature is that they are situated in the centres of population, where thousands of people live crowded together, attracted by available employment and the advantage of town life.

Since dereliction comes principally in the changes of land form and use patterns, the immediate impact is caused on the topography and on the composition of surficial materials- physical, chemical the biological. Most often, a new anthropogenic ecotope is created.

The adverse environmental effects of the derelict lands are: unsightly site topography, site instability, impaired drainage, contamination of air, water and land, and deprivation of ancient rights of the people. Lands become unsustain able.

Most of the derelict land types, sooner or later get colonised by plants different from the old ones except perhaps some of the most inhospigable types. These stressed extreme habitats by being distinctive, definable, often of a known age and ecologically simple, provide opportunities to study the natural colonisers for their success, limitations, adaptive features and evolution.

The study of a derelict land has the principle aim of its reclamation and/or bioremediation. Of all the methods of reclamation, the favoured one is biological, i.e. the use of some kind of plant cover. Natural vegetation that often develops on such sites can be exploited for planning reclamation. Probably, these local native plants have an edge in viability of seeds, ability of self propagation and evolution of tolerant ecotypes. The self attained vegetation cover may indicate species adaptations, and tolerances. Their study may indicate species adaptations and tolerances. Their study may indicate impact of changing habitats on the flora of the general area, loss or rarity of species, invasions of newer ones, indicator status of some of the species to particular substrates or even to recognise some of the old abandoned sites providing refugium to otherwise threatened or rare species.

The original surface of the ground of a derelict land may be changed by terraced excavations, or stripping, subsidence, waste deposites and left over installation. Often a new kind of substate is opened or brought up, strewn or dumped. These substrates appear initially sterile as having no proper soil structure, seed bank, microflora and fauna. They remain unstable, without or very low water holding capacity, most often with severe nutrient shortage- particularly nitrogen and phosphorus, sometimes with adverse pH and presence of heavy metals. Time, climate and growth of natural vegetation reduce some of the adverse factors of derelict lands. The growth of natural vegetation brings about a rapid visual reintegration.

The study of natural vegetation or flora thereof in the following aspects are important : (a) degree of vegetational cover; (b) number of species; (c) species composition (life/growth forms); (d) percentage of woody species; (e) extent and depth of rooting; (f) plant vigour and (g) deficiency/toxicity symptoms (Bradshaw & Chadwick, 1980). According to these authors, site assessement can be made on the following principles : (a) absence or sparseness of colonisers indicate dificiency/toxicity or physical problems of the substrate; (b) a rich floristic diversity may indicate a good potentiality for plant growth; (c) some species can indicate substrate nature; (d) site suitability can be assessed from general appearance and vigour of colonisers and (e) root nature can indicate moisture status of the 'soil'.

As nature might have destined, most of our are minerals occur in tracts once covered by natural forest of one kind and the other almost all kind of industrial development incur alteration of the previous land use and land form. This may involve clearing of forests and existing vegetation as the first step to build the infrastructure and the actual installations. The use of present day technology of heavy earth moving equipments clears the groundm compacts it; bringing in loss in the native natural or semi natural flora, fragmention of forests to remnant patches. The extent in the alteration of the landscape may cover several tens of square kilometers.

On abandonment, absence of vegetation cover in a place may be due to unfavourable rooting medium (compacted ground, rocks, boulders or coarser particles larger than 2 mm in diam.) instability of the surface (looseness of created dumps, acute slopes etc.), low or nil water holding capacity, adverse chemistry and remnant biotic interferences as also inadequate time lapse.

Sparseness of vegetation or patchy distribution of plants may result from any of the causes or their combinations mentioned above, heterogenous nature of the substrate as often they are and colonising abilities of the new recruits.

The habitat conditions in a derelict land area may vary : this may be bare hill slopes, base rocks left over after strip mining, over burden dumps of open cast mining, pit floors and benches of terraced excavations, mine tailings, slag dumps of smelters, pulverized fuel ash and coarse ash of thermal power stations dumped or used as land fills, refuse dumps of various kinds and others.

Depending on the substrate, caesation of activity, climate and availablility of propagules natural colonisation may take place within the first 2-3 years and usually grasses and herbs are the pioneers. This pioneer grass-herb community may thrive for 10 or more years there is a sequential change from the dominance of annuals to biennials and ultimately to a robust perennial *forb* community (Hall, 1957; Bradshaw & Chadwick, 1980; Prach, 1987; Saha *et al.*, 1990). Sporadic occurances of trees can be seen at an early or late stage. As most of such sites are bare at the initial stage, the colonisation that takes place follows a primary successional pattern. According to Prach (1987), total number of plants per unit area increases with age, so is the recruitment of new species but both are limited in time, about 30-50 years.

The advent of colonisers depends primarily on the availability of flora in the surroundings, their variations, propagule types, mode of dispersal, general climate and physico-chemical properties of the substrate. In fact, these are special type of habitat islands.

Anemochorous diaspores principally and to some extent the zoochorous ones contribute to primary colonisation, specially of ruderals. After the establishment of initial colonisation, reproduction by both seed and vegetative spread are the mainstay for the success of vegation development. Grasses are mostly atelo- chorous, i.e. without any special dispersal agent (Saha *et al.* 1990).

Invasion of colonisers starts from the edges (Brierley, 1956). Generally, depressions and bases of slopes are much more convinient for establishment of seedlings and their further growth because of more suitable moisture conditions and accumulation of seeds by water and wind action (Brierley, 1956; Saha *et al.* 1990).

Depending on the physical condition of the substrate, species composition varies as has been seen by Saha *et al.* (1990) in a typical coal field or by Nandi & Mukherjee (1980, 1983) in phosphate and iron mine areas. Ability of plants to become established depends on mechanical rather than chemical factors. Fine materials usually remain free of vegetation as they can not offer any shelter for establishment of seedlings from wind or water erosion. With age, an increase of bulk density in the surface layers (0-5 cm depth) and a decrease in the porosity with depth have been noted by Rimmer (1982) and also a change of pH from basic or neutral to acidic has been noted by Hall (1957). Though, weathering is rock specific, yet some rock types on weathering can support excellent vegetation (Bradshaw & Chadwick, 1980).

Following Russel & La Roi (1986) species can be grouped according to their habitat preferences. Thus, ubiquitous spedies can grow in all kind of places. To name a few they are: *Argemone mexicana, Cleome viscosa, Tephrosia purpurea, Spermacoce hispida, Chromolaena odorata, Tridax procumbens, Xanthium indicum, Hyptis suaveolens, Achyranthes aspera, Croton bonplandianum, Lantana camara, Cynodon dactylon, Euphorbia hirta, Calotropis procera* etc. Then there are xerophytic species commonly growing on coarse textured spoils: *Echinops echinata, Sonchus wightianus, Vernonia cinerea, Solanum erianthum, Peristrophe bicalyclata, Alternanthera pungens, Chloris barbata, Imperata cylindrica Sida rhombifolia, Alysicarpus vaginalis, Elephantopus scaber, Woodfordia fruticosa, Zizyphus oenoplia, Evolvulus alsinoides, Alangium salvifolium* etc. Thirdly, there are species which avoids coarse textured soils: *Sida cordifolia, Corchorus aestuans, Trianthema portulacastrum, Blumea lacera, Eclipta prostrata, Evolvulus nummularius, Leucas aspera, Alternanthera paranychioides, Euphorbia heyneana, Saccharum spontaneum, Thysanolaena maxima*, most of the legumes and rhizomatous grasses. Species growing on rock crevices are few: *Gardenia latifolia, Acacia torta, Adiantum lunulatum, Pteris longifolia, Heminotis ariifolia, Zornia gibbosa Gomphrena celsioides, Paspalum scrobiculatum, Setaria glauca* etc. Species which are common on walls of mine benches are *Canscora decurrens, Hybanthus enneaspermus, Andrographis paniculata, Heteropogon contortus, Chrysopogon aciculatus, Themeda quadrivalvis, Pteris longifolia* etc. On the floors of mine pots are found relatively mesophytic species, common among then are: *Polygala arvensis, Ammania baccifera, Ludwigia perennis, Centella asiatica, Dentella repens, Canscora diffusa, Phyla nodiflora, Murdania nudiflora, Leucas cephalotes, Amischophacelus axillaris, Glinus lotoides, Zeuxine* sp. etc.

From the incomplete list of plants given above, it can be seen that most of them are pantropical weeds. Inspite of site dirrerences, six species - Hyptis suaveolens, Chromolaena odorata, Cleome viscosa, Xanthium indicum, and *Lantana camara* var. *aculeata* are common everywhere (Nandi & Mukherjee, 1980, '83; Sahu *et al.* 1989; Sahu & Santra, 1986; Saha *et al.* 1987).

Total number of species also vary according to the site. Observations by Mukherjee *et al.* (1989) recorded as low as 16 species from refuge dumps of a clay and silicate work in Barakar, to 39 species on thermal power plant ash dump at Jamadoba. The colliery overburden dumps at Chasnala yielded at about 12th year of its existences only 55 species but this increased to 65 species on alluvial dumps at the same place. Sahu *et al.* (1989) recorded about

105 species from coal mine wastes of Talcher and Sahu & Santra (1986) gave a list of 75 species from industrial wastelands of south Bengal, while Saha *et al.* (1990) gave a total of 82 species from Chasnals mines. In the lists of Sahu & Santra (1986) and Saha *et al.*, (1990) most of the plants are dicotyledonous herbs, and Saha *et al.* estimated them to comprise 77.7% of the flora.

The observation by Saha *et al.,* shows that Chamaephytes dominate (29.63%) over other life form classes, followed by Hemicryptophytes (24.07%) and Geophytes (18.25%). Higher number of chamaephytes and hemicrytophytes illustrate the open and discontinuous nature of natural communities, and low number of geophytes may be taken as and indication of comparative immatutity of the therophytes is probably due to the replacement of annuals by perennials. These sites are not specially averse to vegetative growth.

The plant families represented in all these lists show the dominance of families like Fabaceae, Poaceae, As-teraceae, Euphorbiaceae, and Apocynaceae. The narural diminance of grass-legume is a redeeming feature in that they promote stabilisation and nutrient enrichment-particularly nitrogen in all these sites.

Most of the species of the flora of this primary colonisation have the pioneering life form strategies manifested in them. A majority of them are also weedy in nature. The colonisation of open habitats, tolerance of sun and heat, ruderal nature, rapid multiplication/regerenation and no love for vary specific substrate are mostly seen in the ubiquitous species. Inspite of multiple regeneration strategies shown by the dominants (Sahu & Santra, 1986) many of the pioneers have small seeds produced in larger quantities. The small seeds store little amount of reserve food in the endoform or seedling tissue and produce predominantly 'Macaranga' type seedlings (Vogel, 1980). In this type paracotyledons are persistent for a long time and such seedlings can not tolerate shade. At least the six dominant species enlisted beforehand have this type of seedlings (Mukherjee *et al.* 1989).

Root architecture for these species has been observed to belong to the bifurcomonoaxial type (Mukherjee *et al.*, 1989) and in this architectural type more peripherally placed root system has the advantage of utilising subsurface moisture more efficiently.

Germination study in them indicates that they are without any appreciable dormancy and this factor gives these plants the capacity to germinate whenever moisture is not limiting. Moreover, for them the soil pH matches perfectly (Mukherjee, *et al.*, 1989).

The same study has also noted the presence of a number of C_4 spedies among the pioneers. Notable among them are *Amaranthus spinosus, A. viridis, Boerhaavia diffusa, Boreria hispida, Brachiaria ramosa, Chloris barbata, Cynodon dactylon, Cyperus compressus, C. rotundus, Dactyloctaenium aegypticum, Echinocloa cólonum, Euphorbia hirta,* and *E. microphylla*. It is held that the C_4 plants have maximum growth in pre-monsoon times and have higher efficiency in nutrient utilisation in nutrient poor microsites under hot humid climate (Ramakrishnan, 1986).

The stress that are associated with these derelict lands affects some of the morphological traits. The abnormalities are dwarfing of the general stature, development of xeric characters in leaf- particularly glaucousness, size, reduction, reduction or increase in pubescence, anthocyanous colouration, increase in epidermal cell frequency and reduction in palisade ratio, vein islet number and stomatal frequency (Bhairavamurthy, *et al.*, 1985; Kruckeberg, 1984; Saha *et al.*, 1990). The response or kind of abnormality may not be uniform in all the species.

Investigations of derelict lands contaminated with one or more kind of heavy metals is rewarding in a different aspect. More often such areas are colonised by plants that have acquired the capacity to tolerate the toxicity imposed by heavy metals. The screening of species that are capable of direct uptake of contaminants and subsequent acccumulation of nonphytoxic metabolites in plant tissues or which can enhance mineralisation in rhizosphere or release exudates and enzymes that stimulate microbial activity and biochemical transformations can be used in the identification of indicator spacies and their subsequent utilisation in bioremediation of many contaminated derelict lands (Schnoor *et al.*, 1995). There are many reviews dealing with geobotany and indicator plants (s.f. Nayer & Venu, 1990; Meher-Homji, 1989; Tiagi, 1990; etc.). In chromium mining areas at Sukinda, Orissa, Mukherjee (1985) has noted that on mine benches with limonitised wall rocks species like *Canscora decurrens, Hybanthus enneaspermum, Crotolaria albida* and *Andrographis paniculata* are common while on other rock types species common are *Pteris longifolia, Cyperus iria, Chromolaena odorata* and *Andrographis paniculata*. The occurance of *Andrographis panicularta* on both type of rocks is intriguing but plants growing on Ni rich limonites develop chlorosis and/or purplish red colouration of leaves with a veins. Whether this repersented an ecotypic differentiation was not explored further. However, such ecotypic differentiation for *Euphorbia thymifolia, Cynodon dactylon, Melilotus alba* and *Hypericum perforatum* has been reported by Ramakrishnan (1961, 1966, 1968 & 1969) in calcareous habitats. The evolution of edaphic ecotypes has been well established by Snaydon & Bradshaw (1961).

Derelict lands after long period of abandonment can be of exceptional ecological interest. Some of them may turn out of be refugium for a number of rare or threatened species. Gemmell (1977) has listed a number of int esting orchids from alkali waste heaps near Bolton in N.W. Engla id. He has also cited a work by Kelcey (1975) on lime waste bed in Cheshire and which has since been disignated as a nuture reserve. Burt & Bradshaw (1986) have also cited the instance of Millers Dale quarry in the Pek district National Park in U.K.

In age old taxonimic concept, the purpose of a flora is a inventorise the plant wealth of a country and the general trend has been to cover up unexplored or underexplored areas. Widening of database for flora is comparatively a new concept (Heywood, 1984). Dwindling of natural areas and their floras are occuring with the expansion of wastelands. Reclamation of wastelands for whatsoever utilities require vegetation cover for their scenic integration. Natural colonisation though time taking is one option for ecological restoration. An integrated study of the flora using informations on ecology and folristics offer opportunities to assess the impact of changing habitats on the flora, loss or invasions of species, adaptive strategies for pioneers, prospective of some species in phytoremediation and even refugium status of some sites. All such informations are required to select the best option of reclamation.

REFERENCES

Bhairavamurthy, P.V., P.V. Kumar, P. Rethy & Y.V. Anuradha, 1985. *Symp, Biomonitoring State Environ*. 249-253.

Bradshaw, A.D. & M.J. Chadwick, 1980. *Restoration of Land*. Black well Scientific Publishers, London.

Brierley, J.K., 1956. *Transforming our wasteland: The way forward*. Her Majesty's Stationary Office, London.

Gemmell, R.P., 1977. *Colonisation or Industrial Wasteland*. Edward Arnold.

Hall, I.G., 1957. *Jour. Ecol.*. 45: 689-720.

Heywood, V.H. 1984. in Heywood & Moore (eds.) *Current Concepts in Plant Taxonomy* : 397

Kelcey, J.G., 1975. *Environ, Conserv.* 2: 99-108.

Kruckeberg, A.R. 1984. *California Serpentines*. University Cali-, ress.

Mukherjee, P.K. 1985. *Annual Report:* UGC Project "Studies on the correlation of vegetation with copper and nickel ore minerelasation in Bihar and Orrisa" (Unpublished)

Mukherjee, P.K., S.C. Santra, C. Chakraborti & B. Saha, 1989. *Final Report* : DOEN Project "Plant colonisation on Industrial wasteland-survey, ecology and study of biological peculiarities in the reclamation perspectives" (Unpubl.).

Meher-Homji, V.M., 1989. *J. Econ. Tax. Bot.* 13 : 637-654.

Mandi, S.C. & P.K. Mukherjee, 1980. *Bull. Bot. Surv. Ind.* 22 : 63-67

Mandi, S.C. & P.K. Mukherjee, 1983. *Geophytology*. 13 : 145-164.

Nayer, M.P. & P. Venu, 1990. *Bull. Bot. Surv. Ind.* 32 : 28-42.

N.R.S.A. 1993. *Wetland Atlas of India* 1 : Hyderabad.

Prach, K. 1987. *Folia Geobotanica Phytotaxonomica* 22 : 339-354.

Ramakrishnan, P.S. 1961. *J. Indian Bot. Soc.* 40 : 66-81.

Ramakrishnan, P.S. 1966. *Ibid.* 45 : 60-66.

Ramakrishnan, P.S. 1968. *New Phytol.* 67 : 145-157; 301-308.

Ramakrishnan, P.S. 1969. *Can. J. Bot.* 47 : 175-181.

Ramakrishnan, P.S. 1986. *Proc. INSA-British Council Workshop on reclamation of Wasteland*. New Delhi. (Unpubl.)

Rimmer, D.L. 1982. *Jour, Soil Sci.* 33 : 567-579.

Russel, W.B. & G.H. La Roi, 1986. *Can. J. Bot.* 64 : 1286-1298.

Saha, B., C. Chakravorti, P.K. Mukherjee & S.C. Santra, 1987. *Rational Ecological Development*. 123-127. Calcutta.

Saha, B., C. Chakravorti, P.K. Mukherjee & S.C. Santra, 1990a. *Bull Bot. Surv. Ind.* 32 : 86-93.

Saha, B., C. Chakravorti, P.K. Mukherjee & S.C. Santra, 1990b. *Indian Biologist.* 12 : 14-16.

Sahu, A.K. & S.C. Santra, 1986. *J. Econ. Tax. Bot.* 8 : 301-306.

Sahu, R.K., B. Deo, U.C. Mallick & R.C. Maharana, 1989. *Current Science*. 58 : 181-183.

Schnoor, J.L., L.A. Licht, S.C. McCutcheon, N.L. Wolfe & L.H. Carreira, 1995. *Environ. Sci. & Tech.* 29 : 319-323.

Snaydon, R.W. & A.D. Bradshaw, 1961. *New Phytol.* 60 : 219-231.

Tiagi, Y.D. 1990. Sectional Presidential Address; 77th Indian Science Congree. Calcutta.

Vogel, de E.F. 1980. *Seedling of dicotyledons*. Netherlands.

Wallwork, K.L. 1974. *Derelict Land : origins and prospects of a land-use problem*. Newton Abbot.

CHAPTER - 14

KEY ENVIRONMENTAL ISSUES IN SOME INDIAN INDUSTRIES

Arvinda Kumar Jha, S.K. Goel[1] & V. Jha[2]

Department of Botany
L.N. Mithila University, Darbhanga - 846 004
[1] NEERI, Nagpur - 440 020
[2] Deptt. of Biotechnology, LNMU - 846 004

INTRODUCTION

The onset of economic liberalization in the year 1991 has led to unprecedented changes in the Indian industry. The liberal and positive economic measures resulted in manifold increase in the industrial activities, which in turn, brought about a sea change in the overall structure of the industry with big, multinational corporates entering the field and setting up a number of modern production facilities in different parts of the country. This ushered a new era of industrial growth. However, along with the industrial expansion, public awareness on environmental issues have also increased manifold. Environmental issues are now becoming an important factor in shaping the industry's fate. Green lobby is gradually becoming more and more vocal and industries are coming under greater and greater pressure to confirm to various standards of environmental safety and cleaner production as well. While this scenario poses challenges to small and medium sized firms, it also opens new vistas and opportunities, leading to the emergence of environment as a piloting force. In the present communication, key environmental issues facing three major industrial sectors of India, viz., paper, automobile and power are discussed in brief.

PAPER INDUSTRY

As on today, India has 380 paper mills, with the installed capacity of about 44 lakh tonnes. By the year 2005, demand is expected to be 65.7 lakh tonnes comprising 54.8 lakh tonnes of paper and boards and 10.9 lakh tonnes of newsprint. All these will mean that more fibrous raw materials and chemicals will be used, more effluent will be discharged, more carbon dioxide will be released and more Aox will enter the effluent streams (Gopalratnam, 1997).

The key issues of environmental concern associated with the paper industry are :

(a) large use of water and discharge of coloured effluent.

(b) presence of chlorinated by-products in the effluent system.

(c) odour control

(d) acid deposition and stack emission

(e) solid waste

Besides, the industry is also held responsible for the denudation of forests. It is said that one tonne of conventional paper results in the destruction of 277 trees (Kumar, 1995).

The industry has also been indiscriminately discharging its coloured effluents creating aesthetic and chemical pollution of the water bodies. Recently, some attempts have been made, probably with some success, to use the treated effluent for raising agricultural produce, especially sugercane. However, doubts have also been raised on effect of the effluent irrigation of soil condition which can be answered only by scientific study over an extended period of time. Another problem relates to the chlorinated compounds (dioxins) which get into the pulp mill effluent stream through the bleach plant. The negative environmental impact of these chlorinated compounds is well known. Besides, paper industry also results in the emission of sulpher dioxide in enormous quantity which is a source of acid rain. Environmentalists, therefore, desire the paper industry to dispense with chlorine and sulphur and this is something the industry may not be able to comply with fully even in the next ten year's is period.

Strategies : The institute for Environment and Development has made in-depth study on the sustainability of pulp and paper industry. The key concerns may be summarised as follows:

Fibre Sourcing: Replacement of natural forests by plantations; Increased use of non wood fibres and waste papers.

Production: Local production for local needs is desirable; Chlorine use should be replaced/eliminated. and alternative technologies should be investigated.

Consumption: Quality of paper is excessive in many of the uses and rich groups/countries should aim to reduce their consumption of paper as limits of sustainable consumption of paper are exceeded in many cases.

Recycling and waste disposal: Recycled content of most types of paper should be increased; source reduction should be emphasized.

There are numerous other, sometimes conflicting ideas, on the actions necessary to achieve sustainability in the paper cycle. For fibre sourcing, it has been advised to encourage industrial plantations, so that paper industry refrains from sourcing their wood from natural forests. However, many environmentalists are up in arms against industrial plantations due to their being monoculture, thus lacking in biodiversity. Paper industry, however, is a strong supporter of plantations since they are relatively easy to manage, cost effective and more productive than natural forests. The industry is seeking a policy framework for use of degraded forest lands for raising industrial plantations.

In fibre source, recycled fibre is the most favoured source from environmental point of view. Their usage reduces pressure on wood and also results in lower energy consumption, less chemical usage and lower pollutant production.

During production, use of chlorine and its negative impact on the environment has also been a matter of grave concern. The ECF and TCF bleach sequences are favoured by environmentalists. China, India and many other countries of Asia and Africa are yet to switch over to alternative bleaching technologies. Other important issues are : Co-generation, minimising waste usage, reducing emissions, reducing BOD/COD discharge, reducing production of solid wastes and minimising transportation needs.

AUTOMOBILE INDUSTRY

The automobile industry has changed dramatically in the last six years, thanks to the liberazation of the economy. Delicencing of the industry and sharp cut in excise duties have fuelled a quantum growth in production volumes. There have been smart growth in both two-wheeler and four wheeler sectors. In 1993, the production of two-wheelers (Moped, Scooter and Motorcycle) was 17.63 lakh units which rose to 29.74 lakh units in 1997.

In the four wheeler sector, passenger cars are now getting greater prominence. Initially in India, Motor vehicle in general and cars in particular were perceived as luxuries. However, the scenario has now drastically changed. Till the close of the eighties, there were only three to four car manufacturers. However number has now gone upto 13 and the list is still growing.

Until recently, the petrol engine was dominating the passenger car sector and few diesel cars were to be seen. However, the situation has changed dramatically in the last many years and the demand for diesal cars is increasing so rapidly that all the well known car manufacturers are now offering diesal engines as optional fitment on their popular models. In India, where the price of diesel is half than that of petrol, the demand for diesal car, especially in the medium range has forced even the new entrants to offer diesel engine as optional fitment. Maruti Udyog has announced the introduction of diesel model of its zen and Gypsy brands. Telco has also announced the introduction of a new diesel run luxury car named 'Safari', In the three-wheeler sector, Greaves has announced a diesel model 'Garur'. Bajaj auto has also announced a diesel run three-wheeler 'RE-Deisel'.

The production of multi-cylinder diesel engine in the country today is around 6.3 lakh engines a year, with 3.8 lakh units for transport vehicles, 2 lakh for tractors and about 30,000 for construction equipment and genesets. The single-cylinder engines produced are 5.5 lakh units per year. The total for the year 1996, comes roughly to about 1.2 million (Reddy, 1997). It is expected that total production of multicylinder diesel engine may exceed 9 lakh units by the turn of this century and single cylinder engine production may go upto 8 lakh by that time.

The increasing use of diesel over and above that of petrol is a matter a growing environmental concern. It is now a well recognised fact that the particulate matter in the diesel exhaust is a more serious health hazard and strict regulations are being enacted all over the world to reduce the particulate levels in addition to bring down other constituents.

The Indian auto industry may put forward the theory that diesel is comparatively much cheeper and hence vital for economy. However, it is totally disagreable. The Indian diesel is one of the dirtiest in the world. Besides, there is growing concern all over the world regarding toxic emissions of diesel exhaust. Due to government subsidy Indian diesel is the cheapest and hence it is leading to increase in per capita consumption of diesel.

The general perception in India is that diesel passenger car can no longer be detected on the road by noise, smoke or smell. Large improvements have been possible only by reducing sulphur in the fuel and redesign of engines. Reduction of Nox in the presence of oxygen is also attempted now by using DeNox catalyst.

The major pollutions from automobiles are carbon monoxide (CO), hydrocarbons (HC), oxides of nitrogen (Nox), lead compounds, sulpher compounds and particulate matter.

World Scenario of Diesel Vehicles

On worldwide basis, emission control regulations for diesel engines target the control of particulate matter (soot) and Nox. Recent diesel engine modifications have made substantial improvements in the base engine out-emissions of particulate matter (PM) and Nox. In the U.S. a typical heavy duty diesel engine exhaust, a decade ago, was above 0.8 g/BHPH PM and above 6.0 g/BHPH Nox. Engine modifications resulted in a drop to 0.15 g/BHPM PM and 5.0 g/BHPH Nox. These achievements were assisted with a diesel fuel with less than 0.05 percent sulpher content.

To further control diesel engine exhaust emissions, catalytic after treatment devices have been developed. These include diesel oxidation catalyst, a catalytic soot filter and a lean Nox. catalyst. Diesel oxidation catalysts (DOC) are fully developed and are now installed on heavy duty engines in the U.S. since the 1994 model year. DOC catalysts destroy the organic portion of diesel PM. DOCs also destory the gaseous hydrocarbons (HC) which is responsible for the bad odour of diesel engine exhaust. DOCs do not remove the solid carbon soot PM. For several years. DOCs have been installed on passenger car diesel engines in Europe.

The catalytic soot filter (CSF) is in the design and application phase. It is a device that collects all the PM. PM removal efficiency is over 90 per cent. The CSF also destroys the gaseous HCs and co.

The lean Nox catalyst is under fairly intense development phase as a means of diesel engine Nox control. Lean catalyst are not yet commercially available for diesel engine emission control (Govindrajan 1997).

Trends in India

Judging from the present day condition, the quality of Indian diesel is very bad. The high content of sulpher is mainly responsible for the toxic emissions in the exhaust. There is a distinct and linear relationship in between the level of sulpher in diesel and emission of particulate matter.

The presently used Indian diesel has 1 per cent sulpher content. After much hue and cry, the Petroleum Ministry has agreed to reduce sulpher content by 0.05 per cent by the year 2000, whereas present international standards aim at 0.05 per cent sulpher only. Surprisingly, neighbouring country Thailand also uses diesel fuel with 0.05 per cent sulpher only. Sweden has gone much ahead and produces diesel fuel with 0.001 per cent sulpher. Thus, it is evident that even in the year 2000 Indian diesel will be 250 times dirtier than the finest diesel of the world today. And the fact also remains that in no other country the consumption of diesel over and above that of petrol is as high as in India.

Diesel fuel also emits particulate matter in substantive quantity. These particulate matter are encycled with polycyclic aromatic hydrocarbon (PAH), which are carcinogenic. They enter directly into the lungs and are now held responsible for a number of chronic lung disorders especially bronchitis and asthma. According to a 'World Band Report', diesel fuel has the highest per tonne social value. This was concluded in a study conducted on the air pollution and health status survey work relating to various kinds of fuel (Agrawal, 1998).

Besides, one thing that is now gradually understood is that it is not the quantity, rather size of the particulate matter, that makes them more penetrative and health hazard. The submicron particles enter deeper into the lungs and are deposited on the alveolar walls. Till one year before, scientists and environmentalists were worried about PM 10 (Particulate matter, 10). However, now-a-days, matter of more environmental and health concern is PM 2.5. PM 2.5 is of extremely particulate nature, which has the capacity to enter deep into the lungs, trachea and bronchi. Unfortunately the diesel fuel exhaust contains nearly 90 per cent PM 1.

Recent studies on particulate matter emission in automobile exhaust have revealed that the present day world class diesel - those having less than 0.005% sulpher content, also produce 20 times more particulate matter than the petrol. In California, W.H.O, and Air Resources Board have found particulate matter of diesel exhaust highly carcinogenic, states Michal Watas, former director of American Environment Conservation Directorate. In Japan, recently, a polycylic aromatic hydrocarbon (PAH) with nitrate has been discovered, which is said to be most toxic among all the carinogenic substances.

Table 1 shows the evolution of emission standards for two wheelers in India whereas Table 2 shows the evolution of emission standards for passenger cars. The first regulation on pollution norms for vehicles was introduced on April, 1, 1991. All vehicles were expected to meet the specified norms for emission of carbon monoxide and hydrocarbons. Subsequently in 1993, government issued a notification on emission norms to be followed from April. 1, 1996 (Firodia 1995). For meeting the 1997 catalyst norms, unleaded fuel is important and are applicable to cars sold in metropolitan cities only. By the year 2000, unleaded fuel will be available throughout the country. TWC technology with electronic engine management will be the route taken by the auto industry.

Table - 1 : Indian Regulations for two wheelers

Year	Emission standards (g/km)			Comparable stringency
	Co	HC	HC+Nox	
1991	12	8	-	1980 (U.S.)
1996	4.5	-	3.6	1980 (ECE)
2000	2	-	2	EEC (1994)

Table - 2: Indian regulations for Passenger cars.

Year	Emission standards (g/km)			Remarks	Comparable stringency
	Co	HC	Nox		
1991	14.5-27.5	2-2.3	-	Based on weight	ECE 15-04 1982
1996	8.68-12.4	–	3-4.36	Based on engine size	U.S. 1975
1997 Catalyst	4.34-6.2	–	1.5-2.2	" "	ECE 83-1992
1998	Same as 1996, 1997 But with Cold Start		(CAT)	" "	U.S. 1975, and ECE 83-1992
2000	2.72	–	0.97		EEC 1994

At present, standards for PM emission have been set for the year 2001, which are in fact Euro I, that is those regulations which were adopted by Europe in the year 1992. On the other hand, China is adopting Euro I for the year 1996 and Euro II for the year 2001. Besides, the matter of greater concern is that in India, technological advancements regarding toxic emissions of diesel exhaust are in primary stage only. Europe is setting high standards for the year 2000 and 2005 which have been named Europe Phase 3 and Europe Phase 4. Under its proposed standards, the nitrous oxide level of diesel exhaust is three times more than petrol, because, the technology of controlling nitrogen oxide emission has not been developed. Besides, the presence of sulpher in diesel severely limits the utility of catalytic converter.

POWER SECTOR

Thermal Power : In India, still greater emphasis is being laid on thermal source of power without taking into consideration their long term negative impact on the environment. Coal is still the predominant fuel for most power stations and to add fuel, Indian coal has a very high ash content (15-45%).

Today, India has an installed capacity of about 83,288 MW, as on March 31, 1996, against a bare 1362 MW in 1947. The bulk of this commercial power, about 70 percent is produced by fossil fuel combustion, mostly coal. Most of the new additions, unfortunately, are also coal fired thermal power plants. The NTPC alone has so for set up 11 coal based power stations with a total capacity of 13,620 MW and five gas based combined cycle power stations with a capacity of 3175 MW making a total of 16,795 MW throughout the country. New coal based projects with a total capacity of 3000 MW are also being taken up.

After the entry of foreign investers in power generation, the Government of India, since 1991 onwards, received as many as 190 private power project proposals by April 1995 to set up a total generating capacity of 78,260 MW. Of these 147 are thermal projects accounting for 91 percent of the total capacity. This included 74 coal-based projects (N.N.S., 1996).

A coal-based thermal plant pollutes the atmosphere by gaseous emissions of sulpher dioxide, nitrogen oxide etc. A super thermal plant using even normal or low sulpher coal will emit about 100 tonnes of sulpher dioxide a day. The ultimate consequence is acid rain, which damages vegetation, soil and aquatic ecosystem. Besides, tremendous amount of solid waste, flysash and bottom ash are generated. For every megawatt of a thermal plant about one acre of land is required for disposal of the ash generated.

To exemplify, the Talcher thermal plant in Orissa generates 3,000 tonnes of ash a day. When exposed to natural leaching and if only 15 per cent of the toxic metals are leached out, the adjacent Nandira river will receive 208 kg. of iron, 56 kg. of zinc, 45 kg. of copper, 5 kg. of cadmium, 56 kg. of nickel, 4.6 kg. of uranium, 16.5 kg. of thorium, 60.6 kg. of chromium and 11.2 kg. of cobalt, all in a mobile state (Sahu, 1994). All these have a long term impact on the environment. They ultimately find their way into the aquatic ecosystem including groundwater source, resulting ultimately in metal toxicity.

In Orissa itself, a cluster of mega thermal power plants are being set up having a total capacity of 10,000 MW. A statewise campaign has been launched against these thermal plants. Environmentalists are up in arms against these plants due to their negative impact on the environment. It is apprehended that the project would spell serious environmental disaster in the coal belt of the state. It would turn an area of 70 km radius into a man made desert and affect one sixth population of the state with dreaded diseases like cancer, tuberculosis, bronchitis and mental disorder. These plants if set up, it is said, may lead to mass exodus with tonnes of flyash and rise in the atmospheric temperature by nearly 6 degree centrigade in a 70 km radius. An estimated 50 lakh people spread over 10 districts are likely to be hit very badly. Besides, there will be transboarder pollution affecting the adjacent area of Madhya Pradesh. The concern by environmentalists have been corroborated by the recent study conducted by the state Pollution Control Board (Anonymous, 1997). The temptation of immediate revenue should not drive up to actions which lead to future social and environmental disaster. The state will have to bear the brunt of environmental damage, the report observed. The proposed maga power plant sprew spered 2750 kg. of carbon dioxide, 50 kg. sulpher dioxide and 140 kg. of heavy particle along with nitrogen oxide equivalent to fumes exhausted by 15 lakh vehicles per second. Besides, the plant would belch over 2.5 lakh tonnes of ash and throw 2.25 lakh tonnes of flyash into the atmosphere per annum. The apprehension of environmental disaster from the proposed mega plant appears justified and hence instead of mega thermal plants, the government should opt for other sources of energy like hydro and nuclear which are ecofriendly and less hazardous.

Hydropower - India has vast potential of tapping hydel power especially in the Himalayan region. The country is blessed by the monsoons, the snow capped Himalayas and rich river valleys. At present the country has total installed capacity of 2133 MW only. However, the hydropower potential of the country is estimated to be 84,044 MW (Bharadwaj, 1997).

Hydropower is a free gift of nature. It does not cause any type of water or atmospheric pollution. Besides being clean, they are cheep, perennial and inexhaustable. Hydropower stations have a long span, are simple to operate and easy to maintain.

At present the National Hydro Power Corporation, which is the central agency of hydropower generation, is fully conscious of its responsibility towards environmental safety. It is undertaking in-depth works on environmental impact assessment, geo-environmental studies, catchment area treatment, socio-economic studies, study of aquatic eco-system, water quality studies, fish management studies, compensatory afforestation, resettlement and rehabilitation.

Nuclear Power : As compared to other commercial energy sources nuclear power is definitely a cleaner source of power. It has minimal impact on the environment. There is also no submergence of precious land, forest flora and fauna. There are no emissin of toxic gases thus saving the environment from the green house and acid rain effects. If the estimated cost of controlling atmospheric pollution in the case of thermal power is also taken into account, nuclear power will emerge as a clear favorite (Prasad, 1997).

Based on the uranium resources available in the country, it is possible to build, by adopting the fast breeder technology a nuclear power capacity of about 300,000 MW. India had also one of the largest resources of thorium which can also be exploited for the generation of electricity.

Some developed countries which have limited resources of fossil fuel, have fully tapped nuclear energy. Table 3 shows percentage nuclear contribution to total electricity in some of the developed countries.

Table - 3. Contribution of nuclear power in developed countries

Sl. No.	Country	Nuclear contribution to Total Electricity.
1.	U.S.A.	22.49
2.	U.K.	24.99
3.	Germany	29.09
4.	Japan	33.40
5.	South Korea	36.10
6.	Switzerland	39.90
7.	Sweeden	46.61
8.	France	76.14

The total power demand by 2010 is projected at 2,40,000 MW in which contribution of nuclear power is placed at 20,000 MW by the year 2020 (Anonymous, 1998). The present generation of nuclear power in the country has crossed 100 billion units mark. The 100 reacter year of operating experience has been free of any incident and has not led to any release of radioactivity into the environment. Gaseous and liquid releases from the nuclear power stations have been a small percentage of the authorised limits specified by Atomic Energy Regulatory Board.

REFERENCES

Agrawal, A. 1998. Jaher Bujhi Kare; Dainik Hindustan, Patna 19 March.

Anonymous, 1997. Greens up in arms against planned mega power plants in Orissa, The Economic times, Calcutta, 5 November.

Anonymous, 1998. Electricity from Atomic energy; Dainik Hindustan, Patna, 4 June.

Bhardwaj, S.R. 1997. Hydropower: NHPC's mammoth task ahead; SOII, Chennai pp. 67-69.

Firodia, A. 1995. Trends in two wheeler industry. The Economic times, Calcutta 25 May.

Gopalaratnam, N. 1997. Paper clean production, a major issue, SOII, Chennai pp. 411-412.

Govindarajan, S. 1997. Automobile pollution; Emission control gains urgency, SOII, Chennai, pp. 305-309.

Kumar, V. 1995. Recycled hand paper. The Economic Times, Calcutta, 5 June.

N.N.S. 1996. Power Scene crisis and opportunity; SOII, Chennai, pp 59-62.

Prasad, Y.S.R. 1997. Nuclear power : Waiting for a new thrust. SOII, Chennai, pp 79-81.

Reddy, G. 1997. Diesel engines : Sustained rise in demand, SOII, Chennai, page 317.

Sahu, K.C. 1994. Cost of coal combustion; SOE, Chennai, pp 47-51.

CHAPTER - 15

NATURAL DURABILITY OF BAMBOO AGAINST FUNGAL DECOMPOSITION

S. Dutta, P. Dutta Majumdar & N.C. Chatterjee

Department of Botany, Burdwan University
Burdwan - 713 104, West Bengal

INTRODUCTION

Durability refers to ability of any material to resist destructive agencies. The chief destructive agencies, to which the timber is generally exposed, are fungal decay, insect attack etc. In temperate countries, during outside storage, fungal decay brings about the maximum destruction of most of the timber woods than any other cause. On the other hand, in some tropical countries, termites do the most destruction of timbers and as such the term durability is considered to be synonymous with resistance to fungal decay and insect attack in the tropics in particular (Findlay, 1938).

Pulp and paper industries where consumption of bamboos goes annually to 1.85 million tonns approx in our country generally store it in outside locations. As a consequence of storage, ranging from few weeks to about one year, bamboos remain exposed to all the natural hazards. On soil contact, bamboos are attacked by soft rot fungi moulds causing decay, however, the most destructive fungi are Basidiomycetous which attack usually after a few months.

The main chemical constituents of bamboo include cellulose, lignin, hemicellulose, and other extraeous materials like other woods. During the process of decay, the cell wall constituents of bamboo are being degraded and utilized by different decaying fungi.

As a result of such changes, the pulp yield along with its loss of strength and brightness is lost Defiberation is also caused by the attack of white rot fungi. Prominent lysis of cell wall substances as exhibited by the formation of cavities also occurs during decay. Carbohydrate and nitrogenous materials of bamboo woods are also metabolized by the most destructive white rot fungi thereby resulting in the loss of pulp yield and strength.

The fundamental aspects of natural decay resistance of several wood species have been thoroughly reviewed by scheffer and Cowling (1966). Literature are frequently available on natural decay resistance of timber woods but reports on natural decay resistance of bamboo are yet scantly. However, Banerjee and Mukhopadhyay (1962), Guha and Chandra (1979), and Khusniati (1987) carried out some works on the decay resistance of bamboo. Systematic and elaborate studies on this aspect from North-East region of the country where bamboo grows naturally and very luxuriantly in abundance, are lacking. It is expected that the loss of wood substances from bamboo during storage is quite as high as 20-25%. With this aim in view, the present experiment was designed to assess the natural durability of bamboo against fungal decay. Weight loss in a given period of exposure to the test fungi is taken as an inverse measure of decay resistance.

MATERIALS AND METHODS

For the determination of resistance of bamboo against decay by two white rot fungi, both the commonly employed methods such as Agar block method in Kolle flasks and the Soil block method in glass jars with soil as medium were followed (Bakshi *et al.*, 1962-67). The agar block method in Kolle flasks is of the British Standard Method of 1961 and the soil block method is of American Standard Method 1963. From the comparative study, both the methods had been proved to be equally good (Verner and Krause, 1951) but Duncan (1953) reported that the use of soil in soil block method has got some merits over the agar block method because of the fact that (i) soil has the ability to maintain an even moisture content, and (ii) the feeder strips that are used in soil jar method use to supply more natural source of food for the fungal nutrition. Therefore, the accurate determination of the weight loss caused by the test-fungi as a result of decay.

Preparation of test blocks

Three local varieties of bamboo of the age group between 4-5 years (Muli, Bari and Barak) were selected for the tests. For agar block tests small rectangular blocks of uniform size (4.5 x 3.0 x 2.5 cm) for soil jar test were cut from sound bamboo. Uniformly cut blocks were selected for exposure were then subjected to for soaking in distilled water till the moisture content of the individual block remained above the fibre saturation point. The test blocks were freed from any internal fungal mycelium by sterilizing in a moist-atmosphere (Chidester, 1937, 1939). The bibre saturated blocks were then weighed accurately. The average moisture content of the test blocks were noted just before their exposure to the fungal mycelium. The blocks were finally subjected to a condition to temperature at 30°C - 2°C and at R.H. 80%.

Agar block method

For this purpose, the test-fungi were allowed to grow on 2.5% malt-agar medium in Petri dishes which served as the stock cultures of the individual fungus. The malt-extract-agar medium (2.5%) was poured in Kolle flask and sterilized at 15 lbs p.s.i. pressure for 15 minutes. The flasks were inoculated with the test-fungi individually from one week old cultures in Petri dishes and finally subjected to incubation till the fungi covered the entire surface of malt agar.

Transfer of test blocks

The conditioned test blocks were transferred aseptically to Kolle flasks on which the fungal mat was grown. Three test blocks of each host variety were introduced in each flask and finally subjected to incubation for 4, 8 and 12 months.

Soil block method

Cylindrical glass jars of 250 ml capacity fitted with aluminium lids were taken for soil block method. The jars were filled with air dried soil sieved through a filter (filled upto half portion approximately with 150 gm soil/jar). The soil filled jar was moistened with sterile distilled water (150 cc) to each jar. Feeder stripes made from filter paper (Whatman No. 1) of the size (3.5 x 3 cm) were soaked with best nutrient solution for growth of the fungi and placed on the top of the soil in each jar. The jars were then screwed with aluminum lids and sterilized successively for 3 days at 15 lbs pressure.

Inoculation of test-fungi and transfer of test blocks

The test-fungi *Auricularia auricula* and *Dacryopinax*

spathularia were grown on 2.5% malt-agar in Petri dishes uniform discs of fungi mycelia (5 mm dia) were cut from actively growing culture and placed on edge of each nutrient soaked feeder strips. The jars were then incubated at 30°C-2°C and at R.H. 80% till the test organisms cover the feeder strips in the jar. When the fungi covered the feeder stripes, low test blocks of each host variety which were previously oven dried, weighed and conditioned identically as done in case of agar block method, were transferred to the fungal mats with their cut end surface in contact with the fungal mycelium. The lids of the jars were sligh 'y loosened in order to provide aeration to the growing fungi. All he jars were finally incubated at 30C2C and at R.H. 80% for different periods of exposure such as 4, 8 and 12 months. Control sets for both the agar block and soil block methods without the test fungi were maintained identically.

Any change in weight in the control blocks at the end of the experiment which may be due to the factors other than decay was taken for adjusting the weight losses in the test blocks under experiment.

At the end of the incubation periods of 4.8 and 12 months, the test blocks were carefully taken out from Kolle flasks and from glass jars separately and the superficial mycelium adhered to each block was carefully cleaned and weighed to note the moisture content. The blocks were dried at 70°C in an oven to a constant weight. The losses in weight of the test blocks (oven dry weights) were finally recorded in table 1 & 2.

Result and Discussion

It is evident from the results that all the host varieties when exposed to the attack by the test fungi showed weight loses after 4 months. It is also evident that the respective bamboo variety is having a clear resistive capacity against fungal attack. *A. auricula* because of its significant utilization of some of the wood components with the corresponding progress of decay. It is also clear from the result that 'Muli' variety under attack but both the fungi exhibited maximum average weight loss as compared to other two varieties. Of the two methods followed here, Soil block method appeared to be advantageous over the Agar block because of the fact that the former is able to cause decay of the test blocks more naturally than the later which is provided with artificial sources of nutrition.

Table 1. Agar-block method showing the losses in weight of three bamboo varieties after 4, 8 and 12 months of exposure to the attack by *A. auricula* and *D. spathularia*

Host variety	Period of exposure (months)	Average weight loss in percentage*				
		A. auricula	C.D. at 5%	D. spathularia	C.D. at 5%	Without fungus
Muli	4	16.3 (23.81)**	3.61	20.4 (26.85)	2.57	0.0
	8	31.9 (34.39)		35.1 (36.33)		0.0
	12	41.2 (39.93)		48.0 (43.85)		0.5 (4.51)
Barak	4	10.0 (18.43)	2.08	11.4 (19.73)	1.99	0.0
	8	18.4 (25.4)		19.7 (25.41)		0.0
	12	25.5 (30.33)		28.2 (32.1)		0.2 (2.56)
Bari	4	12.3 (20.53)	1.92	14.6 (22.46)	2.04	0.0
	8	22.2 (28.11)		24.3 (29.53)		0.0
	12	27.5 (31.63)		34.5 (35.97)		0.3 (3.41)

* Results are the mean value of five replicates

** Figures in parantheses are average of transformed values = Aresin Percentage

Statistically it has been found that the losses in dry weight of the host woods as result of decay by the test fungi was significantly different during all the time intervals and that too differed significantly from the tissues not exposed to the decay fungi. It may be further concluded from statistical point of view that the results obtained in Soil block method as a result of fungal decay differed significantly higher from.the results recorded in Agar block method.

Table 2. Soil-block method showing the losses in weight of three bamboo varieties after 4, 8 and 12 months of exposure to the attack by *A. auricula* and *D. spathularia*

Host variety	Period of exposure (months)	Average weight loss in percentage*				
		A. auricula	C.D. at 5%	D. spathularia	C.D. at 5%	Without fungus
Muli	4	18.3 (25.32)**	1.96	22.7 (28.45)	2.44	0.0
	8	32.7 (40.97)		38.1 (38.11)		0.0
	12	43.0 (40.97)		49.3 (44.66)		0.3 (3.14)
Barak	4	11.3 (19.64)	2.51	12.5 (20.70)	3.74	0.0
	8	20.0 (26.57)		20.0 (26.57)		0.0
	12	26.1 (30.72)		28.9 (32.52)		0.4 (3.62)
Bari	4	14.2 (22.14)	2.87	15.3 (23.02)	2.51	0.0
	8	25.3 (30.19)		27.0 (31.31)		0.0
	12	30.5 (33.52)		38.1 (38.11)		0.5 (4.51)

* Results are the mean value of five replicates
** Figures in parantheses are average of transformed values = Aresin Percentage

The average moisture content before and after the experimental period was noted which remained much above the fibre saturation point.

From the result it becomes evident that the test blocks of all the host varieties showed losses in dry weight thereby revealing clearly the respective resistive capacity of the host varieties 'Muli' recorded maximum weight loss starting from 4 months onwards as a result of attack by both the fungi whereas the other two varieties showed weight loses below the average loss caused by 'Muli' even after 12 months. Since no single method for field evaluation of decay resistance has been put forwarded, the American Society for Testing and Materials (ASTM, D1758-62) which has got some useful features, has been recommended for wood preservatives (Scheffer and Cowling, 1966). The ASTM standard pertaining to Soil block test for natural decay resistance proved to be highly satisfactory for laboratory evaluations (ASTM, D2017-63). According to this method, weight losses of 0-10, 11- 24, 25-44 and 45% to avove are considered respectively to be the indications of very high, high, moderate and slightly decay resistant.

It is, therefore, clear from the result that 'Muli' variety appeared to be slightly resistant and moderately resistant when exposed to the attack by *D. spathularia* and *A. auricula* respectively at the end of the experimental period. On the other hand, 'Barak' and 'Bari' may be classed as moderately resistant under attack by both the fungi. The difference in relative losses due to decay by the test-fungi in different host varieties taken in the experiment may be constituents such as lignin, cellulose, carbohydrate, nitrogen etc. as exhibited in the previous experiments. The loss in lignin, nitrogen and to some extent in cellulose fractions from 'Muli' existed more than the other two varieties with slight variations among themselves. This may well be correlated with the grouping categories of natural decay resistance where 'Muli' under attack by *D. spathularia* appeared to be slightly resistant and moderate in case of attack by *A. auricula* 'Bari' and 'Barak' are moderately resistant at the end of one year of the experimental period. The loss of wood substances from the host as indicated by the losses of wood substances from the host as indicated by the losses in total weight is evident in the present experiment. Such losses during decay may include many low molecular weight carbohydrates, lignin fractions and extractives (Cowling, 1961). The loss of these substances may either be due to leaching or dur to fungal attack during storage (Guha and Chandra, 1979). Of all the varieties, 'Muli' was found to be either slightly or moderately resistant on the basis of dry weight losses (49.3% and 43.0% respectively) after one year of decay by the test-fungi. This may well be correlated with the presence of comparatively higher level of nitrogen in that variety as observed from the earlier experiment which makes the lost variety vulnerable to infection. The losses in dry weight of the host varieties during decay are in accordance with the results obtained by Banerjee and Mukhopadhyay (1962). Guha and Chandra (1979), Suhiram (1987) in similar type of hosts decayed by Basidomycetes fungi similar results showing losses in dry weight in woody tissues of various hosts during decay by wood-rot fungi have been reported (Morrell and Sexton, 1987, Bellmann, 1988).

REFERENCES

Bakshi, B.K., Puri, Y.N. and S.N. Khan 1962-67. Final technical report on accelaratory laboratory investigations on durability of wood. *For. Path. Br.*, Forest Research Institute, Dehra Dun, India.

Banerjee, S.N. and S. Mukhopadhyay, 1962. A study on *Merulius similis* B & BR and the associated bamboo-rot. *Osterr. Botan. Zeitschrift*, 109(3) : 197-212.

Bellmann, H. 1988. Evaluation of laboratory test results on the resistance of wood species to Basidiomycetes. *Holzals Rohund werkstoff*, 46(8) : 289-294.

Cowling, E.B. 1961. Comparative biochemistry of the decay of sweet gum sapwood by white-rot and brown rot fungi. *USDA Bull.*, 1258 : 79.

Duncan, C.G. 1953. Soil-block and Agar-block techniques for evaluation of oil type wood preservatives. *For. Path. Spc. Relax - US Bur. Ph. India*, No. 37.

Findlay, W.P.K. 1938. The natural resistance to decay of some Empire timbers. *The Empire Forestry Jr.*, 7 : 249-259.

Guha, S.R.D. and A. Chandra, 1979. Studies on the decay of bamboo (*Dacryopinax spathularia*) during outside storage - degradation of cellulose. *Indian Forester*, 6 (June) : 444-450.

Khusniati Suhirman, 1987. Laboratory study on the effect of mud-submerssion treated on the durability of two bamboo and one wood species fungi. *Material and Oraganismen*, 22(4) : 289-296.

Morrell, J.J. and C.M. Sexton, 1987. Decay resistance Port- Oxford cedar. *Forest Products Jr.*, 37(9) : 49-50.

Scheffer, T.C. and E.B. Cowling, 1966. Natural resistance of wood to microbial deterioration. *Ann. Rev. Phytopathol.*, 4 : 147-170.

Verner, R.W. and R.L. Krause, 1951. Agar-block and Soil-block methods for testing wood preservatives. *Indust. Eng. Chem.*, 43 : 1102-1107.

CHAPTER - 16

WASTELAND MANAGEMENT THROUGH SOCIAL FORESTRY IN THE EASTERN BIHAR

T.K. Pan, R.P. Upadhyaya & J. Duttamunshi
TM, Bhagalpur University, Bhagalpur-812 007 (Bihar)

INTRODUCTION

The total geographical area of India encompases approximately 328,778 thousand hectares. India's forest cover according to the latest official estimate has been put to 19.82 per cent of the total land area. As per a recent report of the Forest Survey of India (FSI), the actual forest covers only 64.1 million hectares and that the 80 million hectares is of marginal forests and wasteland. Of the total of 19.82 per cent of green cover forest in India 10 percent is represented by closed forest, 8 percent by open forest and about 0.12 per cent consists of mangrove forest. The green cover also includes 1.10 per cent of the plantation as per report published in The Employment News (May 1998).

So far as the state of Bihar is concerned, the forest covers comprises 2923.2 thousand hectares which is about 17 per cent of the toal geographical area of the state. In Bihar alone about 11 lakh hectares wasteland has been reported (Mandal, 1988). Eastern part of Bihar covers 25 per cent of the green forests. It would be not out of place to mention here that the forest in Bihar is qualitatively very poor. Impact of explosive increase of human population pressure on the forest is indeed phenomenal. This is more particularly remarkable to the forests of Santhal Pargana where sizeable number of tribal population depend directly or indirectly on the forest resources. The local people procure timber, firewood, paper pulp, fibre, fodder and medicine from the surr o} nding forests. forest cover is shrinking and getting thinner due to overexploitation and poor management of the forest. There is yet another turn to the already tight screw. The situation and illimpact is very apparent and has assumed an alarming proportion.

Jhum cultivation is an age-old practice still followed by Santhals and Pahariyas the two ethenic communities of the area causing great damage to the forest and soil profile and loss of genetic resources, some of which may otherwise be very precious. The recent attempts to repair the damage through afforestation under the social forestry has failed to pay the required dividend. This is because artificial forestation can not replace natural one. A natural forest is a well balanced and smooth running ecosystem in itself and any interference as such, further deteriorate the nature and natural surroundings. However, social forestry includes plantations raised in community lands, wastelands, roadsides, canalsides and railway sides and also the plantation raised on degraded forest to repair the ecosystem (Annonymus, 1986).

In Santhal Paragana yet another feature diluting forest cover and disturbing flora is open cast coal mining together with other mining activities which are quite common in this area (Fig. 1). Mining activities are enimical to already depleting and deteriorating floristic composition of the Santhal Paragana region (Varma, 1987). The whole landscape is under reckless unprogrammed and mismanaged industrial activities, generating enormous ecological pressure and causing emergence and expantion of wasteland, which has reached to a figure of 2,38000 hectares.

The alarming situation in this region due to the increase in wasteland formation which is inturn manifested into degradation of socio-economic life of the tribals inspired us to pay proper attention in this area. The wasteland of Santhal Paragana region is ensuing multiple facets some of which are discussed hereunder.

Eastern Bihar's land profile

Eastern boundery of the Bihar state touches West Bengal. Most part of the Eastern Bihar' has uneven land with some exception of Bhagalpur and Munger districts. The whole of Santhal paragana constitute districts of Sahebganj, Godda, Dumka, Pakur and Deoghar, covering hilly tracts which is approximately three eighth of entire area and the rest is alluvial, laterite and hilly ridges

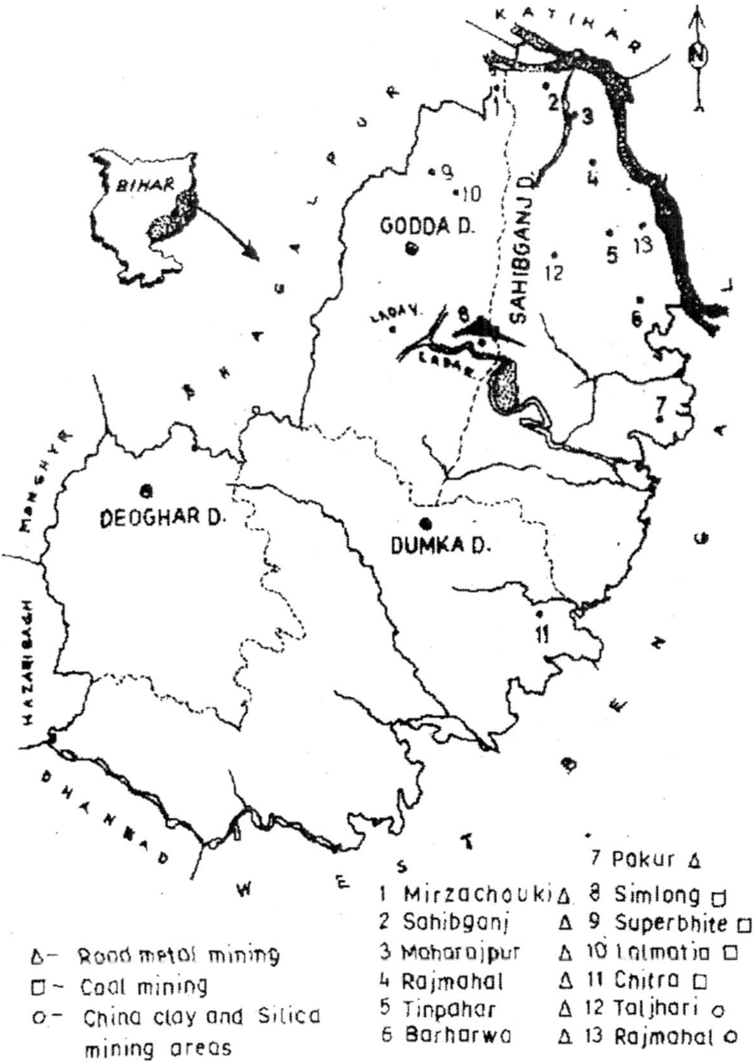

Fig. 1. Map of Santhal Paragana showing major Wasteland Formation Areas under Mining Activities

with Gondowana rocks. Some of the granite rocky areas are completely barren and not suitable for any cultivation. The riverine slopes originated from high mountains having more rich vegetation. The rivers like Gumni, Bonsloi, Brahmani, Ajoy, Lada, Morey are the popular ones along with the mighty Ganga. A low percentage of diara land is also lying along the sides of Ganga bed. Sifting cultivation, mining, tree cutting, overgrazing and construction of dams, roads etc., are the main causes of loss of the top soil of this area (Dutta Munshi, 1990, 1993).

In the Santhal Paragana paradigm culturable wasteland is 7.52 per cent in the Dumka district, 4.31 per cent in the Sahebganj district, 7.01 per cent in the Deoghar district (Datta Munshi, 1989).

WASTELAND TYPES

On the basis of our field experiences wastelands of this area can be divided along the following lines:

(a) Lateritic wasteland

(b) Stony and barren wasteland

(c) Mining area wasteland

(d) Marshy wasteland

(e) Diara wasteland

LATERITIC WASTELAND

A lateritic wasteland is uneven place of land with red morang and small boulders, pebbles such as those found in Jasidih, Madhupur of Deoghar district; Littipara, Kathikund area of Dumka district; Tinpahar, Sahebganj district; Barharwa of Pakur district and also in some parts of Godda district. The soil particles are less in amount in comparison to loamy soil. This area is very suitable for *Shorea robusta* plantation.

Stony and barren wasteland

Santhal Paragana covered with undulating upland with number of hills such as Sundar Pahari, Tinpahar, Panchpahar, Trikut Pahar which is incidentally, the highest peak of Santhal Paragana hills Most of the hills have flat tops where the tribal people cultivate their food grain. Top soil of hilly area is very rich in humus with stony surface underneath. Stony rock without soil cover is completely barren and totally unsuitable for cultivation or plantation.

Mining area wasteland

Santhal Paragana is very rich in mineral resources. Coal mining in Chitra of Dumka district and Siblong, Superbhite, Lalmatia in Godda district and road metal mining in Mirzachowki, Sahebganj proper, Maharajpur, Tinpahar, Barharwa and Pakur, China clay and Silica mining in Taljhari and Rajmahal in Sahebganj district have transformed a good forest area into wasteland. China clay mining of Rajmahal hills is famous since time immemorial. Lalmatia open cast mining is entrenched in about 15 sq. km. besides the big dumping area. Mining activities also run in some other isolated small pockets in the whole of Santhal Paragana region.

Marshy wasteland

Marshy wasteland are formed because of change the river courses. In Rajmahal area of Sahebganj district, marshy wastelands are formed due to construction of Farraka barriage on the river Ganga in 1972. Natural marshy wasteland in Simultala lake (Mirzachowki) and Udhawa lake in Sahebganj district cover a large areas which are confined merely to minor fishing activities.

Diara wasteland

Diara land is formed as a part of the Ganga basin where annual flood is almost a perpetual feature resulting into submergence of land for 2-3 months every year. Submerged land masses along the river bed which emerges out after recession of flood water is called Diara Land. About 1-2% land masses of Sahebganj district is covered by such type of land. Diara is very fertile land because of thick clay sediment which settled down during the flood.

Rescue measure of wasteland

Though social Forestry Programme had been undertaken in First Five Year Plan period in different states of India, but the desired result is not obtained because of some lacuna and lapses. First and foremost is noninvolvement of local mass in its meaningful true sense for the success. It is the local population who is the first beneficiaries of the whole idea and this very fact has not been regulated in the past excercises. Further, in earlier schemes depleted forest land has not been undertaken as the target. Mostly projects such as Sweedish International Agency (SIDA) and World Bank have been working in this field. In Bihar specially eastern part has great potential and of-late forest department has undertaken thickening of forest area as a reclaimation measure. Further selection of plant for the sake of social forestry has not been very scientific earlier, as such the desired benefit and ecological repair could not be made possible. Enlisting of suitable plant species has been done in Table 1 which could meet the objective of social forestry. The recommendations of the National Commission on Agriculture (1976) are the followings :

Table 1. Recommended plants of wasteland development

TaxonName with common name	Suitable wasteland for plantation	Uses
Acacia auriculae formis Australian babool	All type of wasteland	Wood & fuel
Acacia catechu Khair	Stony & mining	Katha & fuel
Acacia nilotica Babul	Diara & laterite	Fodder & Timber
Aegle marmelos Bel	Laterite	Fruit
Albizzia lebbeck Sirish laterite	Marshy, mining, stony	Timer & fodder
A. procera Sada sirish	Do	Do
Alstonia scholaris Chatman	Laterite & Stony	Timber & medicinal
Anacardium occidentale Kaju	Do	Fruit & fuel
Annona squamosa Sarifa	Laterite, mining & stony	Fruit & fule
Artocarpus integrifolia Jackfruit, Kathal	Mining, laterite & Stony	Timber & fruit
Azadirachta indica	Mining, laterite	Wood & medicine
Bambusa arundinacea	Mining, laterite	Timber & Fuel
Barringtonia acutangula	Marshy	Timber & Fuel
Cassia saimea	Mining & Stony	Timber & Flower
Casuarina equisetifolia	Mining, Stony & laterite	Timber & soil binding
Ceiba pentandra Green simul	Mining & stony	Timber & cotton
Dalbergia latifolia Black sisham	Mining, stony & & laterite	Timber & fuel

D. sissoo Sisham	Do	Do
Delonix regia	Do	Timber & flowring
Dendrocalamus strictus	Mining, stony &laterite	Fruit
Embilica officinalis Amla		
Dillenia pentagyna	Mining, stony & Laterite	Fruit
Eucalyptus sp.	All type of wasteland	Eucalyptus oil, fuel wood & Timber
Ficus bengalensis Bat, Bar	Mining, stony & laterite	Timber & fodder
F. religiosa Peepal	Mining, stony & laterite	Fodder & fuel
F. glomaruta Gulhar	Mining & laterite	Timber & fodder
Gmelina arborea Gambhar	Diara and laterite	Timber & fuel
Grevillea robusta	Mining &literite	Ornamental & Timber
Lagerstroemia speciosa Jarul	Stony & Marshy	Ornamental & Timber
Madhuca indica Mahua	Laterite & Stony	Fruit & Timber
Mangifera indica Am	Laterite, stony & mining	Fruit & Timber
Mimusops elengi Maulsiri	Laterite	Medicine, Fruit & Timber
Moringa oleifera Sajina	Stony & laterate	Vegetable
Pongamia pinnata Kanji	Marshy, stony & Mining	Timber
Salmalia malabarica Semul	Laterite & stony	Fuel wood & cotton
Sesbania grandiflora Agasth	Mining & Literite	Vegetable & fuel wood
Shorea robusta Sal	Laterite	Timber, fuel wood medicine
Sterculia urens	Mining & Stony	Timber
Syzygium cumini Jamun	All types of wasteland	Fruit Medicine & Timber
Tamarix dioca Jhao	Marshy land	Timber & fuel
Tectona grandis Teak, Segaun	Stony & laterite	Timber

Terminalia arjuna	Marshy & laterite	Medicine,Timber & Tasar culture
T. alata Assan	Do	Do
T. bellirica Bahare	Do	Do
T. chebula Harre	Do	Do
T. catappa	Do	Do

(A) Lateritic wasteland

The land is very suitable for sal (*Shorea robusta*) plantation. Soil consists of pebbles and cobbles and has very low percentage of NPK and further has a pH range of 6.6-7.0. This type of wasteland is found all around the Santhal Paragana. Agro-forestry practices are most suitable in such a type of soil.

(B) Stony and Barren wasteland

The whole of Santhal Paragana region is upland hilly tract. Most of the area are covered with scrub forest except Selingi and Kathikund area in this region. On hill slopes and hill tops age-shifting cultivation should be replaced by scientific method. Agro-forestry and forest based industry producing plants are suitable for social forestry plantation with active participation of the tribals. Suitable crops for this purpose are Makai (*Zea mays*), Bajra (*Pennisetum typhoides*), Bora (*Vigna sinensis*), Kokori (*Momordica dioica*), Marua (*Eleusine caracana*), Arhar (*Cajnus cajan*), Sabai grass (*Eulaliopsis binata*), Zinger (*Curcuma species*) and also suitable plants for agro-forestry are bamboo (*Bambusa* sp.), Teak (*Tectona grandis*) Jamun (*Syzygium cumini*), Sal (*Shorea robusta*) etc.

(C) Mining area wasteland

Open cast mining generate vast area of barren land where top soil is completely destroyed and has zero level of NPK and with various pollutent materials. The proper testing of soil and suitable type of plant species are yet to be selected for plantation and rehabilitation of land. Under such a category Australian babool (*Acacia auriculaeformis*), *Sesbania* spp. *Casuarina equisetifolia*, *Cassia* spp. may be selected as are most suitable species.

(D) Marshy wasteland

Marshy wasteland is waterlogged area where plantation may be done in the fringe areas of water beds. Jamun (*Syzygium cumini*), Arjun (*Terminalia arjuna*), Jarul (*Lagerstroemia speciosa*), Sissoo (*Dalbergia sissoo*) etc. are only waterlogged tolerant species suitable for plantation in such an area.

(E) Diara wasteland

Diara land is river side land which remain inundated every year during rainy season by flood and emerge-out after the rainy season when water recedes. This is ideal for the cash crop cultivation during winter and summer months. Very limited species are suitable for afforestation purpose in this area which include Jhowa (*Tamarix dioica*), Jamun (*Syzygium cumini*), Arjuna (*Terminalia arjuna*), Babul (*Acacia nilotica*), Gambhar (*Gmelina arborea*) etc.

For Agro-forestry - *Thysanolaena maxima, Eulaliopsis binata, Momordica dioica, Vetivaria* Sp. *Curcuma* sp., *Morus alba* are suitable species to cultivate in stony, laterite and mining wasteland as cash crops.

Having gone through available literature and by our own personel experience of decades the above mentioned species were recommended for plantation in different categories of wasteland.

REFERENCES

Anonymous 1986. Social forestry in India Problems and prospect, Birla Institute of Scientific Research, *Economic Research Division*, Barakhanda Road, New Delhi -1, Page 35.

DuttaMunshi, J.S., S.P. Roy and J. Dutta Munshi 1989. Ethnozoological survey and ecology of the tribal area of Santhal Parganas (Bihar), *Final Technical Report*, Department of Environment, Govt. of India, New Delhi.

Dutta Munshi, J. and C.B. Singh 1990. Shifting cultivation in Santhal Parganas, Bihar, *Environment and Ecology* 8(2) : 679-682.

Dutta Munshi, J. *et al.*, 1993. Natural Resources of Santhal Parganas, *Environment and Ecology*, 11(1) : 80-83.

Mandal, R.B. 1987. Wasteland development of Bihar. *Proceeding Seminar State of Environment*, Bhagalpur University, Page 17.

Varma S.K. and A.K. Pandey 1987. Vanising living and fossil flora of Rajmahal Hills. *Pros. Seminar State of Environment*, Bhagalpur University, pp. 46-55.

CHAPTER - 17

STANDARDISATION OF NURSERY TECHNOLOGIES : ESTABLISHMENT OF GERM PLASM BANK AND MASS PROPAGATION OF FUEL WOOD SPECIES

S.C. Pal

P. I. Biomass Research Centre & Professor of Botany
Visva Bharti, Santiniketan - 731 235, W.B.

INTRODUCTION

MNES sponsored Biomass Research Centres have been established at different agro climatic zone of the country. Biomass Research Centre at Visva Bharti has been established for biomass plantations at its research field with a view to identify suitable fuel wood species for this region. The main objective of the present study is to select proper nursery site, soil mixture, proper time of seed collection and their viability, proper storage and to undertake germination studies to determine water and nutrient requirement of seedlings and to determine their proper size and age for transplantation collection of suitable genotypes and phenotypes and selection of genetically superior trees for mass propagation of superior planting materials.

MATERIALS AND METHODS

The following fuel wood species were alloted for the present investigation namely *Acacia auriculiformis, Alstonia scholaris,*

Termanalia arjuna, Bauhinia variegata, Azadirachta indic, Butea monosperma and *Madhuca latifolia.* The site lies in the central part of subhumid lateritic belt of West Bengal with a temperature variation of 43°C (max) and 9°C (min) and a normal annual rainfall of 1285 mm (average of a year).

Predominent soil type of these area is red and lateritic and old aluvial. The soil is acidic (pH 4.5 to 7.1), content of organic carbon, available P_2O_5 and K_2O are 0.11-1.7%, 30.0-117.8 and 30-500 respectively. A systematic survey on the selected fuel wood species in the District of Birbhum was carried out. Phenological studies of these species were also conducted. Two nursery beds were raised, one at Santiniketan and other at Dubrajpur with total area of 14000sq. feet each. Seeds were sown both in the bed and poly packs for quality seedlings/saplings. Each species was studied separately in terms of their age, productivity and total biomass. Water requirement, age and size of seedlings for trans- plantation we are ascertained and standardised. Flowering time, fruit and seed setting, dormancy, germination and growth of these recommended species were studied. Seeds were collected from superior trees for provenance trial. Phenotypically superior plants of each species were taken into account in terms of their age, productivity and total biomass.

RESULTS AND DISCUSSION

A monthly schedule of different nursery activities is presented in the Table 1.

Table 1. Monthly Schedule of Different Nursery Activities

MONTH	ACTIVITIES
October	Selection of nursery site
November	Ploughing and cleaning of nursery site
December	Preparation of mother bed, preparation of irrigation and drainage channel.
January	(a) Preparation of soil, manure mixture and filling of polythene tubes. (b) Seed collection for *Acacia auriculiformis, A.holosericea, A. mangium, Terminalia arjuna, Dalbergia sissoo.*
February	(a) Filling of polythene packets completed. (b) Seed collection for *A.auriculiformis, T.arjuna, Cassia fistula, C.siamea*

March	(a)	Collection of *Acacia auriculiformis, Alstonia scholaris, T. arjuna, C. fistula, C. siamea.*
	(b)	Sowing of seeds of : *A. auriculiformis, C. fistula, C.siamea, Dalbergia sissoo.*
April	(a)	Collection of seeds of *Bauhinia veriegata, Butea monosperma.*
	(b)	Sowing of seeds of : *B. variegata, Butea monosperma, A. scholaris, D. sissoo.*
	(c)	Irrigation of nursery.
May	(a)	Collection of seeds of : *Azadirachta indica, B. variegata, B. monosperma.*
	(b)	Sowing of seeds of : *A.indica,*
	(c)	Irrigation of nursery.
	(d)	Weeding and hoeing of nursery beds and polythene
	(e)	Shifting of polythene packets
June	(a)	Collection of seed of : *M. latifolia, Artocarpus heterophyllus, A. indica.*
	(b)	Sowing of seeds : *M. latifolia, A.hetetophyllus, A.indica.*
	(c)	Regular irrigation of nursery.
	(d)	Weeding and hoeing of nursery beds and polythene packets.
July	(a)	Irrigation of nursery.
	(b)	shifting of seedlings.
	(c)	Transplantation of seedlings which are ready for transplantation.
August	(a)	Transplantation of seedlings in the field.
Sept.	(a)	Reorganisation of the nursery.
	(b)	Shifting of residual seedlings.
	(c)	Cleaning of vacant nursery beds.

The results of viability, dormacy, germination and seed storage studies of each species are given in the Table 2. Which shows that dormancy of seed varies from 30 days to 60 days, germination from 87% to 96% and Viability varies from 2 months to 1 year. Table 3 show water requirement, age and size of seedling for transplantation. It is evident from the Table 3 that most of the species required water once daily except *Acacia auriculiformis* and *Alstonia scholaris* which required thrice daily. Age and size of seedling for trans-plantations is roughly 2.5 to 3 months and 20 to 30 cm respectively in most cases. Biomass characters of different fuel wood species at the age of 1.5 years is represented in Table 4. From the Table 4 it is quite evident that total biomass is much higher in 1m x 1m spacing for all the species in comparison to other two spacing. As our objective is to produce more fuel wood in less amount of land and as there is provision of periodical coppicing, plantation can be recommended (1, 4). So far as the total biomass *per ha.* is concerned *Bauhinia variegata* is found to be best followed by (in descending order) *A. auriculiformis T. arjuna A. scolaris M. latifolia B. variegata*

Table 2. Viability, Storage, Dormancy and Germination percentage of selected fuel wood species.

Species	Dormancy (%)	Germination (%)	Storage	Viability
Acacia auriculiformis	60 days	5	Room tem.	1 year
Alstomia scholaris	60 days	96	"	3 months
Azadirachta indica	Nil	97	12-15°C	7 days
Bauhinia variegata	30 days	96	12-15°C	6 months
Butea monosperma	30 days	87	12-15°C	2 months
Madhuca latifolia	Nil	85	12-15°C	1/2 year
Terminalia arjuna	70 days	93	12-15°C	1/2 year

Table 3. Water requirement, age and size of seedling for transplantation.

Species	Water req. of seedling	Age of seedling	Size (length) of seedling (in cm)
Acacia auriculiformis	first 15 days twice daily then once a day	2.5-3 months	30-40 cm
Alstomia scholaris	First 25 days twice daily then once a day	3 months	30-40 cm
Azadirachta indica	First 7 days thrice daily then once a day	3-3.5 months	40-45 cm
Bauhinia variegata	Once a day	2.5-3 months	40-50 cm
Butea monosperma	Once a day	3 months	20-30 cm
Madhuca latifolia	Once a day	2.5-3 months	20-25 cm
Terminalia arjuna	Once a day	2.5-3 months	45-50 cm

Table 4. Biomass Yield Characters of Different Fuel Wood Species At the Age of 1.5 years

Species		Heigh (cm)	Girth (cm)	Total Biomass (kg) at diff. spacing					
				1m x 1m		1.5m x 1.5m		2m x 2m	
				AG	UG	AG	UG	AG	UG
A.auriculiformis	(R)	195.8	08.9	2.3	0.75	2.34	0.79	2.4	0.85
	(S)	201.7	9.05	2.4	0.90	2.37	0.80	2.4	0.84
Termnalia arjuna	(R)	96.66	10.0	0.6	0.49	0.61	0.59	1.3	1.14
	(S)	110.9	10.4	0.7	0.57	0.63	0.59	1.4	1.24
Madhuca latifolia	(R)	53.0	7.9	0.16	0.23	0.17	0.24	0.2	0.26
	(S)	59.0	8.0	0.16	0.22	0.17	0.25	0.2	0.25
Azadirachta indica	(R)	99.8	6.7	0.20	0.13	0.22	0.15	0.2	0.15
	(S)	110.6	7.0	0.20	0.13	0.24	0.16	0.3	0.17
Alstonia scholaris	(R)	102.9	10.5	0.61	0.33	0.63	0.36	0.7	0.42
	(S)	107.8	11.3	0.62	0.35	0.63	0.37	0.7	0.45
Bauhini variegata	(R)	175.6	7.9	2.25	1.17	2.39	1.19	2.4	1.20
	(S)	183.6	8.6	2.25	1.10	2.45	1.20	2.5	1.35
Butea monosperma	(R)	29.7	4.0	0.05	0.30	0.05	0.31	.05	0.37
	(S)	35.5	4.5	0.04	0.30	0.05	0.35	.05	0.40

NOTE: Mean weight of 10 plants of each species were considered for each set.
R = Taken from Random Sample.
S = Taken from Selected Sample.
AG = Stands for Above Ground biomass
UG = Stands for Under Ground biomass.

From the evaluation studies it has been established that the seedlings raised from selected plants were superior than the seedlings raised from the seeds collected at random from group of plants. A germplasm bank for different fuel wood species for provenance testing is established and maintained.

The results of comparative studies of growth character of different fuel wood species show that the biomass yield character of the seedling differ in the following order: *A. auriculiformis T. arjuna B. variegata A. indica M. latifolia A. scholaris* A procedure has been standardised for selection of plants as superior seed source for mass propagation. Techniques for mass propagation seedlings through seeds have been standardised. The following package of practice have been developed for the tree species which are as follows:

Package of practices developed for five species out of alloted ten species, are as follows:

I. *Alstonia scholaris*

1.	Botanical Name	:	*Alstonia scholaris*
2.	English Name	:	Devil tree
3.	Common/local Name	:	Chattim
4.	Family	:	Apocynaceae
5.	Indegenous/Exotic	:	Indigenous
6.	Flowering time	:	September to October
7.	Fruiting time	:	March to April
8.	Description of the plant	:	Trees; leaves 3-4 whorld; flowers in subterminal, corymbose cymes; sepal-5; petal-5, connate; stamens adnate near apex of tube; carple-2, distinct; fruit of 2 linear, slender follicles; seed oblong, linear; taproot with branches, coppicing capacity - good.
9.	Seed Production	:	Profuse
10.	Method of seed Storage	:	Dormancy 60 days; viability 90 days; storage at room temperature (30-35°C).
11.	Seed germination treatment	:	No special treatment is required
12.	Propagation	:	Seed propagation.
13.	Method of nursery raising	:	Media of seed germination is soil (Well pulverised) and cowdung in 4:1 ratio; time of seed sowing - April to May.
14.	Transplanting period	:	At the age of three months
15.	Manuring	:	Mixture of soil and cowdung manure is the ratio 4:1; foliar spray of 2% urea solution at the age of 15 and 30 days after germination.
16.	Irrigation	:	Watering once daily throughout the nursery period.

II. *Butea monosperma*

1.	Botanical Name	:	*Butea monosperma*

2.	Common/localName	:	Palash
3.	Family	:	Fabaceae
4.	Indegenous/Ecotic	:	Indigenous
5.	Flowering time	:	February to April
6.	Fruiting time	:	April to May
7.	Description of the plant	:	Trees/leaves pinnately 3-foliolate; flowers large sohwy in racemes; fruits are oblong. Seed obovate, compressed; tap root with branches.
8.	Seed Production	:	Profuse
9.	Method of seed storage	:	Dormancy 30 days, viability 60 days, storage at room temperature.
10.	Seed germination treatment	:	Soaking of seeds in lukewarm water for 12 hours.
11.	Propagation	:	Seed propagation
12.	Method of nursery raising	:	Medium of seed germination is soil (Well pulverised) and cowdung in 4:1 ratio; time of seed sowing May-June.
13.	Transplanting period	:	At the age of three months.
14.	Manuring	:	Mixture of soil and cowdung manure in the ratio 4:1; foliar spray of 2% urea solution at the age of 15 and 30 days after germination.
15.	Irrigation	:	Watering once daily throughout the nursery period.

III. *Madhuca latifolia*

1.	Botanical Name	:	*Madhuca latifolia*
2.	English Name	:	Mahua
3.	Common/localName	:	Mahua
4.	Family	:	Sapotaceae
5.	Indegenous/Exotic	:	Indigenous
6.	Flowering time	:	March to April
7.	Fruiting time	:	May to June
8.	Description of the plant	:	Trees; leave petiolate, coriaceous, silky or tomentose beneath when young, clustered at ends of branches; fruit oblong berry; seed ellipsoid.
9.	Seed Production	:	Profuse

10.	Method of seed Storage	:	Dormancy 7 days, viability 7 to 10 days; storage at room temperature (30-35°C) or in low temperature.
11.	Seed germination treatment	:	No special treatment is required
12.	Propagation	:	Seed propagation
13.	Method of nursery raising	:	Soil, sand and cowdung mixture at the ratio 2:2:1; time of seed sowing is June to July.
14.	Transplanting period	:	At the age of 2.5 to 3 months.
15.	Manuring	:	Mixture of soil and cowdung manure is the ratio 4:1; foliar spray of 2% urea solution at the age of 15 and 30 days after germination; better response by applying VAM fertilizer have been recorded.
16.	Irrigation	:	Watering once daily throughout the nursery period.

IV. *Terminalia arjuna*

1.	Botanical Name	:	*Terminalia arjuna*
2.	Common/local Name	:	Arjun
3.	Family	:	Combretaceae
4.	Indigenous/Exotic	:	Indigenous
5.	Flowering time	:	April to May
6.	Fruiting time	:	July to April
7.	Description of the plant	:	Large tree; leaves with short petiole, elliptic in old trees, lanceolate in seedlings; fruit ovoid, coriaceous drupe with hard indehiscent endocarp, externally with five angles or wings; wing of fruit marked with much ascending striations; seed solitary; taproot with branches; coppicing capacity good.
8.	Seed Production	:	Profuse
9.	Method of seed Storage	:	Dormancy 70 days, viability 1.5 years; storage at room temperature.
10.	Seed germination treatment	:	Soaking of seeds in cold water for one month before germination is required.
11.	Propagation	:	Seed propagation

12.	Method of nursery raising	:	Medium of seed germination is soil, sand and cowdung at the ratio of 2:2:1; time of seed sowing June to July.
13.	Transplanting period	:	At the age of 2.5 to 3 months.
14.	Manuring	:	Mixture of soil and cowdung manure is the ratio 4:1; foliar spary of 2% urea solution at the age of 15 and 30 days after germination; better response by applying have been recorded.
15.	Irrigation	:	Watering once daily throughout the nursery period.

V. *Bauhinia variegata*

1.	Botanical Name	:	*Bauhinia variegata*
2.	Common/local Name	:	Kanchan
3.	Family	:	Fabaceae
4.	Indegenous/Exotic	:	Indigenous
5.	Flowering time	:	February to March
6.	Fruiting time	:	March to April
7.	Description of the plant	:	Medium sized tree, flowers vary a good deal in colour (various shades of violet to purple). Fruit an oblong linear pod, seeds ovate, top root with branches.
8.	Seed Production	:	Profuse
9.	Method of seed testing & Storage	:	Dormancy 30 days, viability 6 months; storage at room temperature.
10.	Seed germination treatment	:	Soaking of seeds in water (12- 15°C) for 12-18 hours.
11.	Propagation	:	Seed propagation
12.	Method of nursery raising	:	Medium of seed germination is soil, (well pulverised) and cowdung at the ratio of 4:1; time of seed sowing is during April.
13.	Transplanting period	:	At the age of 2.5 to 3 months.
14.	Manuring	:	Soil and cowdung manure is the ratio of 4:1; foliar spary of 2% urea solution at the age of 15 days and at 30 days after germination; better response have been recorded by applying VAM biofertilizer.
15.	Irrigation	:	Watering once daily throughout the nursery.

VI. *Acacia auriculiformis*

1.	Botanical Name	:	*Acacia auriculiformis*
2.	Common/local Name	:	Sonajhuri
3.	Family	:	Fabaceae/Mimosaceae
4.	Indigenous/Exotic	:	Exotic
5.	Flowering time	:	July to November
6.	Fruiting time	:	November to December
7.	Description of the plant	:	Tree; leaves reduced to phyllodes, curved, coriaceous, tapering at the end, parallel-nerved confluent near the base; flowers small, yellow, in cylindrical axillary spikes, fruit-pods moniliform, coiled, dehiscent. In case of seedlings bipinnate compound leaves present.
8.	Seed Production	:	Profuse
9.	Method of seed testing & Storage	:	Dormancy 60 days, viability 1 years; storage at room temperature.
10.	Seed germination treatment	:	Soaking of seeds in luke warm water (12-15°C) for 12 hours.
11.	Propagation	:	Seed propagation
12.	Method of nursery raising	:	Medium of seed germination is soil, (well pulverised)and cowdung at the ratio of 4:1; seeds are sown either in seed bed directly or in poly bags during April to May.
13.	Transplanting period	:	At the age of 2.5 to 3 months.
14.	Manuring	:	Soil and cowdung manure is the ratio of 4:1; foliar spary of 2% urea at the age of 15 days and at 30 days after germination; better response have been recorded by applying VAM biofertilizer.
15.	Irrigation	:	Watering once daily throughout the nursery.

VII. *Azadirachata indica*

1.	Botanical Name	:	*Azadirachata indica*
2.	Common/local Name	:	Neem
3.	Family	:	Meliaceae
4.	Indegenous/Exotic	:	Indigenous

5.	Flowering time	: March to April
6.	Fruiting time	: May to June
7.	Description of the plant	: Tree; tap root with branches; leaves alternate pinnate subopposite; flowers calyx-5, connate at base, petals-5, free spathulare, stamen-10 fruit drupe, oblong, yellow; seed ellipsoid, solitary;
8.	Seed Production	: Profuse
9.	Method of seed Testing & Storage	: Dormancy nil, viability 7 days; storage at room temperature.
10.	Seed germination treatment	: No such treatment is required.
11.	Propagation	: Seed propagation
12.	Method of nursery raising	: Medium of seed germination is soil, (well pulverised)and cowdung at the ratio of 4:1; seedlings may be raised bed directly/ poly bags; time of seed sowing is June-July.
13.	Transplanting period	: At the age of 3 months.
14.	Manuring	: Soil and cowdung manure is the ratio of 4:1; foliar spary of 2% urea at the age of 15 days and at 30 days after germi nation; response have been recorded by applying VAM biofertilizer.
15.	Irrigation	: Watering once daily throughout the nursery.

The results proved that selected plants of known provenance possessing superior biomass yielding characters produced seedlings having greater height and stouter stem.

A procedure has been standardised for selection of plants as superior seed source for mass propagation. A germplasm bank for different fuel wood tree species has been established and maintained and a techniques for mass propagation of seedlings through seeds have been standardised.

REFERENCES

Bandyopadhaya, D. 1986. "Integrating forestry with rural development" workshop on waste land development, N. Delhi, Feb. 4.

Kaul, R.N. and B.N. Ganguly 1963. studies on the economics of raising nursery seedlings in arid zone. Ann. of Arid zone I(2): 85-105.

Kushalappa, K.A. 1986. Waste land development techniques. National seminar on F.W. and F.P. from wasteland. May 30-31, Dehradun.

Pandey, S.C. 1987. Waste land development-need for holistic approach, 16th Oct, Economic Times.

CHAPTER - 18

MANAGEMENT OF SOAPSTONE MINE WASTE DUMPS WITH SPECIAL REFERENCE TO ITS RECLAMATION

Anal K. Jha & B.C. Mishra*

University Department of Chemistry
T.M. Bhagalpur University, Bhagalpur - 812 007, Bihar
* Chief Technical Adviser, Golcha Group of Industries, "Prem Prakash", S.M.S. Highway, Jaipur.

The process of reforming and regarding the wasteland approximately to its original topography and to bring about permanent self sustaining vegetation is called reclamation. It is indeed, an integral part of mining activity. In surface mining, the first layer of the overburden to be removed is the top soil (thick or skeletal), this gets burried deep inside the dumps and foible soil consisting mainly of sand boulders, pebbles and devoids of organic matter and nutrients having very poor or virtually no moisture retaintion capacity occupy the upper layers.

Thus, ecological balance of the soil and flora is disturbed in a jiffy. Nature may take hundreds of years to restore the same. The exploitation of mineral wealth creates a series of new wastelands in the form of minewaste dumps.

Mining of soapstone generates wastes in large amounts as the O.B. to mineral ratio is generally in a range of 17 : 1 (or it may go up to 44 : 1 at times), waste dumps chiefly constitute boulders of quartzite, low grade dolomites and at times low grade soapstone. However, reclamation of these waste dumps can involve the following three steps :

(1) Land scaping or site preparation

(2) Soil accumulation, and

(3) Revegetaion

(1) Land Scaping

This covers all the activities used for removal of soils and O.B., disposal of wastes and modefications of disturbed land alongwith waste disposal sites in order to ensure a complete reclamation of the mined area. But in order to have special emphasis on reclamation of the waste dumps (old) of the soapstone, one can go for afforestation only. Site preparation is carried out with a view to improve the physical condition of the overburden for eventual land use. Old dumps are first graded to a gentle slop of less than 1:4 or 5 (Fig. 1). The graded surface is then covered with 15-30 cm. thick layer of top soil to ensure proper growth of vegetation.

When reclamation is taken side by side with the mining operation, the reclamation activities cannot wait till the mining is complete. Under such conditions, construction of new tips can be undertaken.

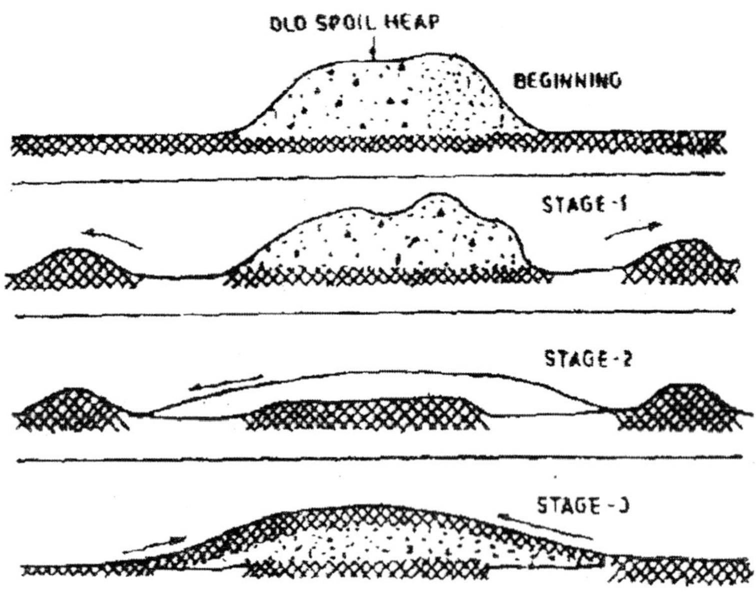

Fig. 1. Regarding an Old TIP, Stripping and Replacing Topsoil

Top soil and sub soil should be analysed for their capacity to support vegetation and if they are found to be of better quality, as it generally happens with the soapstone, then the material should be tipped and stacked properly. This stacking of top soil should not be done for a longer period as it may lose viability. Sequence of activities of constructing a new tip on plane grounds with an aim of future reclamation is shown in Fig. 2. Progressive reclamation with revegetation is followed as the tipping continues. Similarly perimeter tipping which consists of constructing the outer tip faces first and then infilling the intermediate area can also be followed both on the level ground and stopping ground as shown in Fig. 3. The outer faces of the perimeter tips are covered with good soil and the revegetation is started simultaneously. The progressive reclamation allows maximum time for vegetation to grow.

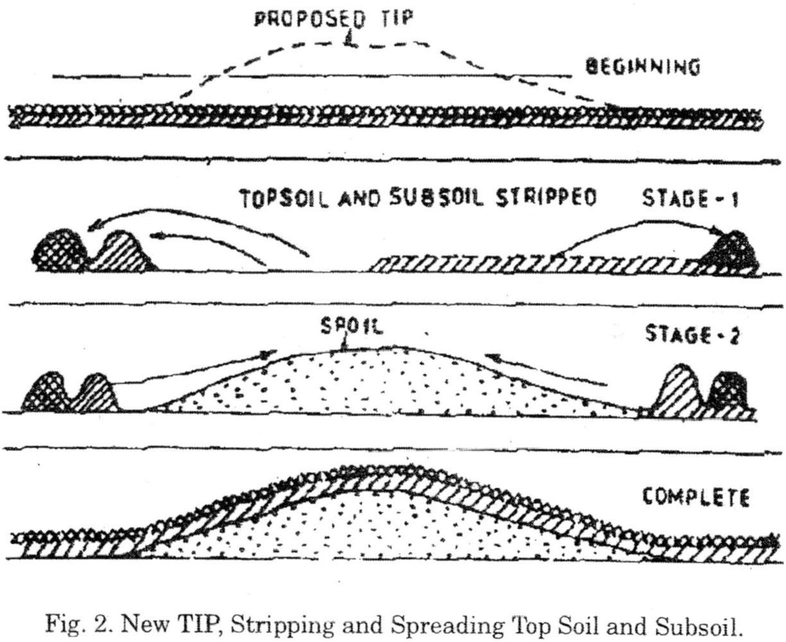

Fig. 2. New TIP, Stripping and Spreading Top Soil and Subsoil.

It is very important to consider slop of the site to be reclaimed vis-a-vis the ultimate use of the surface area. Gradients are made keeping in view the slope stability of the material. Grading is generally done by the help of bull dozers or loaders. Maximum slope satisfactory for afforestation is 35°. If reclamation is undertaken in an arid zone like Rajasthan, pitting, gouging and land imprinting is of great help.

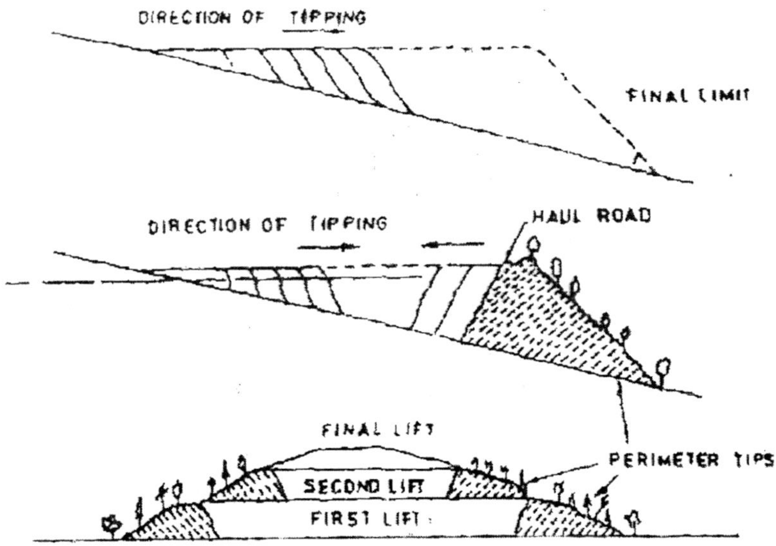

Fig. 3. Showing Process of Perimeter Tipping During Early Stage of Mining Activity.

2. Soil Accumulation

Soil amelioration activities are undertaken to overcome the adverse conditions including salinity, pH, deficiencies of nutrients, poor texture and structure of soil and lack of organic matter. Soil amelioration is necessary to commence the revegetation. Most of the mine wastes require some amount of soil amelioration prior to the revegetation programme and the soapstone mine waste dumps are no exception.

Soil accumulation for revegetation consists of (i) Laying or spreading an additional layer of top soil as required, (ii) treatment of the surface soil with mulches or adhesive chemical stabilisers (iii) soil amendment if required.

In case of soapstone mine waste dumps, neither mulching nor soil amendment is required because of methodical dumping of heavier boulders at the bottom in order to achieve better stability as well as due to the neutral nature of the mineral. Sometimes, Ammonium nitrate based blasted material used for plantation over waste dumps have given quite encouraging results in terms of plant growth.

3. Revegetation

The vegetation cover spread over any mine spoil/area is beneficial in following manners :

(i) Stabilising erodible slopes to minimise pollution of adjoing areas.

(ii) Control of dust

(iii) Enhancement of aesthetic value

(iv) To maximise evo-transpiration which helps in minimizing run off.

(v) Fascilitate productive land use for agro-forestry.

(vi) Reducing oxidation which often gives rise to acid mine drainage (not in case of soapstone)

(vii) Wide belts of forest trees help in reducing noise.

Studies have shown that forests are very effective in controlling erosion. Forests also reduce ambient temperature of area. Afforestation programme at Neyveli Lignite Corporation Ltd., has caused lowering of average temperature of the area from 42°C to 39°C and helped to increase the average annual rainfall from 1190 to 1500 mm.

Revegetation can be attempted in two different ways. Either the plant species suitable for the existing soil conditions can be considered for revegetation, or the soil and site conditions can be made suitable to facilitate the growth of such plant species. However, with soapstone mine waste dumps, native species of the area like Babool, Khejri, Neem, Amaltas, Moringa, Australian acacia etc. have been found to be quite effective. On the slopes of the dumps, during initial stages, Solanum janthocarpum, Ipomea fistulosa and a few grass species come up quite nicely. Plantation of Australian acacia is particularly recommended for the soapstone mine waste dumps because these plants are not only fast growing, rather they also have nodules in their root system which facilitates fixation of atmospheric Nitrogen. Recent VAM inoculation technology and Vermiculture could also be of great help.

REFERENCES

Banerjee, S.P. 1987. Land Reclamation in Mined areas. In National workshop on Environment Management of Mining operations in India. B.H.U., Varanasi.

Chandana, C. 1995. Blasting and Plant growth. In souvenir of the 5th Mine Environment and Mineral Conservation week Open cast Mines-Ajmer region 1994-95.

Choudhary, J.R. 1995. Reclamation of Mined area. In the Souvenir of Mine Environment and Mineral conservation week; Open cast Mines Ajmer region 1994-95.

Coppin, N.J. and Bradshaw, A.D. 1982. Quarry Reclamation, Mining Journal Books, London.

Environmental planning of mining industries through forestry. In Environmental management of Mining operations. Department of Environment, Ministry of Environment and Forests, Paryavaran Bhawan, CGO complex, Lodhi Road, New Delhi.

Hanan, J.C. 1984. Mine Rehabilitation, New South wales coal Association, Sydney.

Marshall, I.B. 1983. Mining, Land use and the Environment, II, A review of mine reclamation, Activities in Canada. Land use in Canada Series 23, Environment Canada, Ottawa.

Marshall, I.B. 1982. Mining land use, and the Environment land use in Canada series no. 22, Canadian Overview, Environment Canada, Ottawa.

Williamson, N.A., Jhonson, M.S. and Bradshaw, A.D. 1982. Mine waste Reclamation. Mining Journal Books, London.

CHAPTER - 19

ECORESTORATION OF MINED WASTELAND : AN OVERVIEW

Ambarish Mukherjee & Debnath Palit

Department of Botany, Burdwan University
Golapbag, Burdwan - 713 104, West Bengal

AN OVERVIEW

World's economic progress depends to a large extent on exploitation of minerals, knowing fully well that mining is detrimental to our environment. The ever-increasing involvement in extraction of ores for fast developing industries has been ravaging the life sustaining systems to an alarming extent. The situation, being very grave, demands earnest endeavor to encounter hazards emanating from stripping or opencast mining, compaction, tailings, acidity, metal toxicity and dissemination of pollutants from the ore-dependent industries. High mineral content and toxicity of the mined and mine adnate areas prove hostile to plants and turn the area into a wasteland with spoiled soil (spoil). In juvenile mining areas vegetation proceeds gradually towards total devastation with the progress of mining. Often restoration of original biodiversity on once mined area seems impracticable. Damages wrought by concomitant pollution are of no insignificance and the effects on the ecosystems, human life, ground water, aquifers etc. necessiate much careful attention. There is deterioration of agriculture, horticulture, forestry, pasturing etc. The global land area perturbed every year by mining is about 386,000 hectares. In India mining affects about 683,672 hectares of land dispersed in 19 of her states. We are myopic to the huge burden of spoiles and pollutants emanating therefrom and the work on environmental restoration is virtually inadequate.

Recently ecorestorative work on iron-ore, bauxite, lignite and limestone spoils have been started in India. Programmes for revegetation of limestone and rock phosphate mining areas in Dehradun, bauxite, dolomite and coal overburden in Madhya Pradesh, opencast mine spoils of lignite at Neyveli, coal spoils of Talcher, Dhanbad, Raniganj etc. have been started and the measures so far taken are upto satisfaction of environmentalist (Mukherjee, 1993b).

The Government insists upon reclamation of devastated surfaces before sanctioning new leases for mining. Certain legislations should be enacted to monitor the derangements resulting from mining and mandate storage of top soil for reuse, refilling of mined areas, revegetation and restoration of natural hydrology.

Lands turn barren while coming in menacing contact of mining operations and allied works. The vegetation emaciates in areas unable to evade from metal toxicity. Once favourable soils in the premining periods get transformed into "spoils" through inductive alteration in physico-chemical properties and impoverishment of nutrients. The low moisture content of spoils inhibit plant growth considerably. Spoils often have a lower infiltration and water holding capacity (Pederson *et al.*, 1980) and a higher bulk density (Smith *et al.*, 1971) compared to those of unmined soils.

Spoil-banks consist of heterogeneous mixture of overburdened strata and their composition and texture have a profound influence on the physical properties found at the surface of the spoil. Erosion is less severe in case of the surface soil (spoil) containing a significant amount of coarse fragments (Smith *et al.*, 1974). Spoils containing as much as 75 per cent coarse fragments (by weight) do not seriously reduce productivity.

Acidity of spoils is often the prime factor determining establishment of vegetation on spoil material (Tyner and Smith, 1945). Establishment of vegetation becomes impossible when the value of pH falls below 4. For such areas grasses like *Cyanodon dactylon, Panicum clandestinum, Echinocola crusgalii* etc. can be used since they can withstand very low pH. Stucky *et al.* (1980) have reported that roots of *Phalaris arundinacea, Panicum virgatum* and *Dectylis glomerata* can withstand pH values of 3.5, 3.7 and 4.2 respectively. Acidic mine spoils support a very poor number of bacteria, actinomycetes and fungi (Wilson, 1965; Lawry, 1977) so that there is a highly reduced biolocial phenomena manifested in slow decomposition of litter and recycling of nutrients. Besides rendering acidity, chemical weathering of toxic spoils often produces high concentration of soluble salts and their eventual mobilisation. The findings of Vimmerstedt and Struthers (1968) established that most toxic spoil-banks improve with time in regard to establishing vegetation. However, erosion on toxic spoil-banks removes the surface materials optimised in course of time and exposes fresh toxic materials to the plants.

An important component of a favourable rooting matrix for plants is organic matter as it has a desirable effect on soil tilth, nutrient supply, water infiltration and retention. The organic matter content of spoils is considerably lower than that of natural soils (Smith *et al.*, 1971). Nitrogen is the chief limiting factor for plant growth (Mays and Bungston, 1978). A legume grass association if engaged in ecorestoration can not retain top soil but also eliminate the need for application of nitrogen fertilisers to maintain vegetation (Woodruff and Blaser, 1970).

Ecorestoration of mined lands is very intricate, complicated and site and climate specific work. Several methods may be adopted for ameliorating the mined soils viz. application of liming materials, sewage sludge, paper mill sludge, power plant fly ash ore and covering of spoils with fertile soil material brought from elsewhere so that plant growth can be established. For a vegetation cover the spoil may require alteration of both physical and chemical properties since they usually have detrimental pH and texture, deficiency of nutrients and moisture besides many toxic substances. Furthermore, a thorough analysis of environmental constraints, threat perception and meteorological status is necessary to steer amendments towards ecorestoration. To develop a healthy environment in mined areas large scale optimisation strategies should be designed hand in hand with wind screening measures since erosion is a major hindrance that not only reduces ecological productivity but also causes off site damage by depositing toxic sediments in roadside ditches, water courses, water reservoires and productive agricultural land in the vicinity. The success of revegetation depends on the selection of proper species which should be based on physico-chemical properties of the substratum, topographic influence, climate, use and management planned. The plants should be selected from those which existed in the premining period or those tolerant to site specific problems. There are some plant genotypes which have a high constancy and fidelity values on soils overloaded with a particular metal due to their narrowly specialised adaptation. These plants are called metallophytes. They have biomass with abnormally high metal content in correlation with high amount of metal in their habitats and are capable of stabilising soil and man-oveuring conditions for biosuitable environment. Many studies have been conducted to identify metallophytes complementary to specific metals in the substratum (Tiagi and Aery, 1985) and to determine adaptations of various plant species for revegetation and stabilisation of mined lands (Tyner and Smith, 1945; Grandt and Lang, 1958; Vogel and Berg, 1968; Bennett, 1971; Brown, 1971; Ruffner, 1978). Work on identification of metallophytes in India is pioneered by Venkatesh (1964) and propagated by Gandhi and Aswatha Narayan (1975, Avery (1977), Tiagi and Aery (1981, 1985). These work dealt mainly with identification of specific metallophytes for geobotanical exploration of underground resources (Mukherjee, 1993a) and did not attempt to put their findings in reclamation of mined land. However, these work should be honoured with decisive significance to revegetate as much spoil as possible. The metallophytes that can be considered suitable for ecorestoration of mines include grasses, legumes, trees and shrubs. Copper tolerance of *Agrostis tenuis*, cadmium tolerance of *Impatiens balsamina*, zinc tolerance of *Anthoxanthum odoratum*, lead tolerance of *Festuca rubra*, cobalt tolerance of *Silene cobalticola* and *Crotalaria cobalticola* etc. are known (Sharma, 1981). *Rumex actosella* (Polygona case) is a metallophyte that can tolerate aliminium and accumulate relatively high concentration of iron, magnesium and manganese. *Artiplex nummularia* grows well on nickel-rich spoils and *Artiplex canescens* on Uranium and coal spoils. Other members of Chenopodiaceae can also be tried on soils rich in soluble salts. *Robinia pseudoacacia* is very suitable for highly acidic soils and are resistant to lime, magnesium, lead, zinc, sulphur, iron, coal and spoil-banks of fire clay and brick wastes. Many grasses and sedges can grow well in and around coal and allied mines. Plants like *Ceratophyllum* and *Vallisnaria* can accumulate heavy metals (Mukherjee, 1993b). In metal contaminated wetlands *Eichornia crassipes, Juncellus inundatus*, species of *Cyperus, Typha, Carex* and others may be used. These are useful particularly along the edges of drainage of tailing dams.

Grasses, legumes, non-leguminous nitrogen fixing plants (e.g. *Alnus, Casuarina, Elaegnus, Trema* etc.) and ectomycorrhizal trees can serve as pioneers in revegetation programme. These plants can improve the site for further succession through their root system and contribution of organic water. Many species of *Eucalyptus* (*E. camaldulensis, E. citridora, E. microtheca* and *E. tereticornis*) can be grown on mine spoils. *Araucaria cunninghami, Khaya senegalensis, Pinus sylvestris, Swietenia macrophylla* are known to improve rock phosphate soils.

On tailings which have soluble salts certain xerophytic or halophytic plants may be tried since they can use deeply seated water table and resist high salinity. The use of algae, lichens and mosses in stabilization of tailings and other overburdens can prepare a very favourable matrix for establishment of higher plants. Inoculating spoils with suitable microfloral (bacteria, actinomycetes, fungi) is one of the most effective techniques since it can drive recycling of nutrients through decomposition. Introducing earthworms, termites and soil dwelling insects into spoil can augment incorporation or organic matter, porosity and finally revegetation.

There are other ways of reclaiming mined lands. Clothing of spoils with top soil can be adopted for establishing vegetation (Sutton, 1979). An exotic fertile soil superimposed on spoil can afford accommodation to a wide range of non-metallophytes. Further more, penetration of their roots into underlying spoil can develop firm integrity to subdue the possibility of the cover being removed by sliding.

Mulches act as a soil protectant and enhance plant establishment (Kay, 1978). Mulching with agricultural refuse, pulp fibre, sawdust, straw, woodchips, cured hay etc. can change the micro- climate governing the spoil, the surface temperature of which is often high and detrimental. Mulches trap sediments and protect the site by reducing impact of raindrops, soil erosion and water run-off and increasing water infiltration. Mulches retain fertilizers and seeds in place, conserve moisture for seed germination, protect seedlings, prevent crusting modifying temperature of the soil. The most commonly used materials are organic substances such as, residues from agricultural and wood processing industries (Vogel, 1981). Myer *et al.* (1972) reported that crushed stone, gravel and woodchip mulches have great potential for erosion control on steep barren slopes, especially where conventional methods might not be satisfactory.

Power plant fly ashes when used in accurate quantities can promote revegetation on toxic spoils (Plank and Martens, 1973; Martens and Beahm, 1976). However, severe toxicity symptoms in plants were recorded when large amounts of fly ash were applied to strip mine areas (Bennett, 1971).

Large quantities of sewage sludge have also been used for

initiating vegetation on spoils (Hill *et al.*, 1979; Montague and Hill, 1976; Sutton, 1979; Haghiri and Sutton, 1982). A sludge by- product of the paper mill industry has been examined as an amendment for establishing vegetation on spoils and has given excellent results (Hotinik and Watson, 1982; Haghiri and Sutton, 1982). Paper mill sludges contain large amounts of basic cations and plant nutrients. They also contain high levels of readily decomposable organic matter which can be beneficial in altering the physical and chemical properties of spoils but have an inhibitory effect on available nitrogen (Dolar *et al.*, 1972).

Since mining is essential for economic development this should not be put in confrontation with conservation. It should be in vogue keeping harmony with ecorestoration programmes. Reclamation of mined areas and optimisation of their environment should be undertaken at a war-foot basis together with more rationally planned mining policies and ore-based industrial practices so as to keep economic development unimpaired. Of the different measures botanical method of reclamation is the best one since once a vegetation is established the aesthetic value and productivity of the area is increased. It is likely to have a long term success is making them available for future landuse and economic return.

REFERENCES

Aery, N.C. 1977. *Geobios* 4 : 225-228.

Bennett, O.L. 1971. In : "Revegetation and Economic use of surface mined land and mine refuse symposium" (D.M. Bondurant, ed.), pp. 23-25. School of Mines, West Virginia Univ., Morgantown.

Brown, J.H. 1971. In : Revegetation and Economic Use of surface mined land and mine refuse symposium (D.M. Bondurant, ed.) pp. 26-28. West Virginia Univ., Morgantown.

Dolar, S.G., J.R. Boyle, and D.R. Keeny, 1972. *J. Environ. Qual.* 1 : 405-409.

Gandhi, S.M. and U. Aswathanarayana, 1975. *J. Geochem. Explor.* 4 : 247-250.

Grandt, A.F. and A.L. Lang, 1958. Reclaiming Illinois Strip Coal Land with Grasses and Legumes. Univ. of Illinois Agric. Exp. Stn. Bull. 628, Urbana.

Haghiri, F. and P. Sutton, 1982. In : Land Reclamation and Biomass Production with Municipal Wastewater and Sludge (W.E. Sopper, E.M. Speaker and R.K. Bastian eds.), pp. 433-446. Pennsylvania State Univ. Press, University Park, Pa.

Hill, R.D., K.R. Hinkle, and R.S. Klingensmith, In : Utilisation of Municipal Sewage Effluent and Sludge on Forest and Disturbed Land (W.E. Sopper and S.N. Kett, eds.), pp. 425-443. Pennsylvania State Univ. Press, University Park, Pa.

Hotinik, H.A.J. and M.E. Watson, 1982. In : Land Reclamation and Biomass Production with Municipal Wastewater and Sludge" (W.E. Sopper, E.M. Speaker and R.K. Bastian eds.), pp. 301-306. Pennsylvania State Univ. Press, University Park, Pa.

Kay, B.L. 1978. In : Reclamation of Drastically Disturbed Land (F.W. Schaller and P. Sutton, eds.), pp. 467-483. *Am. Soc. of Agron.* Madison, Wisconsin.

Lawry, J.D. 1977. *Environ. Pollut.* 14 : 195-205.

Malyuga, D.P. 1964. Biogeochemical methods of prospecting. Consultant Bureau, New York.

Martens, D.C. and B.R. Beahm, 1976. In : International Ash Utilisation Symposium, pp. 657-664. Morgantown Energy Research Centre, Morgantown, West Virginia.

Mays, D.A. and G.W. Bengston, 1978. In : Reclamation of Drastically Disturbed Lands (F.W. Schaller and P. Sutton, eds.), pp. 307-327. *Am. Soc. Agron.*, Madison, Wisconsin.

Montague, A. and R.D. Hill, 1976. The potential for using Sewage Sludge and Compost in Mine Reclamation. Mining Pollution Control Report. *Ind. Environ. Res. Lab.*, Cincinati, Ohio.

Mukherjee, A. 1993a. *J. of Sane the Environment* 11(2) : 9-12.

Mukherjee, A. 1993b. *Yojana* 37(22) : 17-18.

Myer, L.D., C.B. Johnson, and G.R. Foster, 1972. *J. Soil Water Conserv.* 27 : 264-269.

Pederson, T.A., A.S. Rogowski, and R.Jr. Pennock, 1980. *Soil Sci. Soc. Am. J.* 44 : 321-328.

Plank, C.O. and Martens, D.C. 1973. *J. Soil Water Conserv.* 28: 177-179.

Ruffner, J.D. 1978. Plant Performance on Surface Coal Mine Spoil in Eastern United States, USDA Soil Conserv. Service SCS-TP-155. Morgantown, West Virginia.

Sharma, A.K. 1981. Impact of the Development of Science and TEchnology on Environment. Presidential address 68th Session Indian Science Congress Association, Varanasi, pp. 1-43.

Smith, R.M., E.H. Tyron, and Tyner, E.H. 1971. Soil Development on Mine Spoil. *West Virginia Univ. Agric. Exp. Stn. Bull.* 604T, Morgantown.

Smith, R.M., W., Grube, T. Arkle, and Sobeck, A. 1974. Mine Spoil Potentials for Soil and Water Quality. EPA 670/2-74-070. U.S. Environmental Protection Agency, Cincinnati, Ohio.

Soil Survey Staff, 1966. Soil Survey Manual, USDA Handbook No. 18.

Stucky, D.J., J.H. Bauer, and T.C. Lindsey, 1980. *Reclamation Rev.* 3 : 129-139.

Sutton, P. 1979. *Ohio Rep.* 64 : 72-75.

Sutton, P. and W.A. Dick, 1987. In : Advances in Agronomy (N.C. Brady ed.), pp. 377-405. A.P. Inc., San Diego, California.

Tiagi, Y.D. and N.C. Aery, 1981. *Proc. Indian Natn. Sci. Acad.*, 47(6) : 867-887.

Tiagi, Y.D. and N.C. Aery, 1985. *Symp. Biomonitoring State Environ.*, pp. 207-222.

Tyner, E.H. and R.M. Smith, 1945. *Soil Sci. Soc. Am. Proc.* 10: 429-436.

Venkatesh, V. 1964. *Ind. Min.* 18 : 1-101.

Vimmerstedt, J.P. and Struthers, P.H. 1968. In : Second Symposium on Coal Mine Drainage Research, pp. 152-163. Bituminous Coal Research, Monroeville, PA.

Vogel, W.G. 1981. A Guide for Revegetating Coal Mine-soils in the Eastern United States. Gen. Tech. Rep. NE-68. Northeastern Forest Experiment Station, Broomall, PA.

Vegel, W.G. and Berg, W.A. 1968. *J. Soil Water Conserv*. 23 : 89- 91.

Wilson, H.A. 1965. The Microbiology of the Strip-Mine Spoil. W. Va. Univ. Agric. Exp. Stn. Bull. 506 T, Morgantown.

Woodruff, J.M. and R.E. Blaser, 1970. Establishing and Maintaining Turf on Steep Slopes along Virginia Highways. Annual Report Virginia Dep. of Highway and U.S. Bureau of Public Roads, Dep. of Agronomy, Virginia Polytechnic and State Univ., Blacksburg.

CHAPTER - 20

MANAGEMENT OF WASTELAND WITH STAGNANT WATER

K.R. Maurya & B.B. Mishra

Bihar Agricultural College, Sabour,
Bhagalpur, Bihar, 813 210.

INTRODUCTION

Land degradation causing deterioration of its potential and productive values due to reasons like water and wind erosion (150 Mha), salinity and alkalinity (7 Mha), waterlogging (6 Mha), shifting cultivation (3 Mha) and ravines, riverine and torrents (7 Mha) has been documented in all the 15 agroclimatic zones of India (Kadrekar 1993). Water conservation is, by and large, the prerequisite for soil conservation. India annually receives 400 Mha meter water, of which 70 Mha metre gets evaporated, 115 Mha metre is runoff and 215 Mha metre gets infiltrated in the sub-soil state (Kadrekar 1993). However, some man-made artificial wastelands do also occur along the highways, railway tracks, mining sites in depression which are loaded with rain water for major period of year and are left uncultivated due to adverse physical, chemical and biological conditions.

The stagnation of water during rainy season is a good natural resources to be exploited purposefully for developing a land use system. The objective is to develope the waste land for best profitable biomass generation.

However, such degraded waste land has never been taken into consideration either by the Government or by private or public agencies. These lands have widely been exploited through forestation by the Deptt. of Social Forestry mainly. The present paper aims at proper utilization of this particular type of waste land by harvesting the rainwater and growing profitable water loving crops like *Makhana* and *Singhara*.

Present Scenario

These waste lands are occurring in a narrow strip in both sides of highways and railways track (about 50 m). Even if these roads and railways tracks, for example, cover a minimum total distance of 1 lakh Km throughout the country, the waste land will roughly be 50 thousand hectare. This is far less than the actual acreage, since other sectors like industrial area, mining site etc do also occupy such waste land in India.

Interestingly these degraded area has never been monitored even through the satellite imageries and till date, it is practically neglected. No survey work has been launched to delineate the area for this mission. Obviously immediate attention is required towards such problematic zones which are potentially productive but practically waste land.

Water harvesting vis-a-vis soil and water conservation

Water harvesting in general refers to a method of storing water in deep surface ponds or in deep underground storage tanks or as underground water through percolation system made artificially (Singh and Mishra, 1996). However, in the present context, water harvesting aims at storing rainwater in deep surface ponds with minimum loss so as to help the water loving crops sufficiently during the off rainy seasons.

Conservation of either soil or water or both in this context is an integrated approach of transferring barren or wasteland into productive one and thus conservation in such man-made waste land is not merely an anti-erosion or anti-runoff measure but it is to build up resources for optimal production in a given socio- economic environment. For this, landuse planning system should be developed which will surely include forest trees, water loving plants and agricultural crops. The sloping portion of the roads/railways track will be made covered with forest trees in both sides whereas areas under depression will be re-shaped by conservation having deep ponds followed by levelled cropland. The gradient of shoping portion, size and depth of ponds as well as levelled crop land will depend on climatic data as well as resurces appraisal (Singh and

Mishra, 1996) so that profitable return could be obtained on minimum investment. Our knowledge about forest plantation and crop cultivation is well emphasized. However, awareness to grow water loving crops like *Makhana* and *Singhara (Paniphal)* after constructing ponds in such specified location is restricted. These two crops are highly remunerative and can be cultivated on large scale.

Preference for growing water loving crops

Makhana and *Singhara (Paniphal)* are two important water loving fruit crops which may successfully be grown in ponds constructed in planned manner in both sides of the roadways and railways tracks.

1. MAKHANA

Makhana (Eurale ferox) belongs to Nympheacea family with broad leaves full of thorns. Thorny morphology restricts fishes to grow in ponds. Soils and water qualities of Darbhanga, Madhubani, Purnea, Katihar and Saharsa of north Bihar are suitable for *Makhana* cultivation. About 90 percent production of *Makhana* in India is recorded in Bihar and 75 percent of total production in Bihar is estimated mostly in 5 districts of north Bihar.

Sowing time of *Makhana* seed falls between November and December. On germination, each plant has 15-20 branches. Flowers appear on the water surface and gradually go inside water on fruiting obviously due to weight. On maturity of fruits, pedicle and leaves start decaying leaving behind folicle which is actual fruits. Each plant has 10-15 folicle and each folicle has 100-200 seeds.

Makhana is a nutritive and digestible dry fruit and its flakes contain 9.7 percent protein, 76.9 percent carbohydrate, 0.1 percent fat, 1.3 percent salt and 11.29 percent moisture. *Makhana* is a remedy for iron in human body, since it contains 1.4 per cent iron in it.

The most limiting factor for growing *Makhana* is the consistent water depth of pond minimum being 60-75 cm. The area must have heavy textured soils rich in organic matter with slightly acidic pH (5.0-5-6). Ponds must be free from weeds.

No recommended varieties of *Makhana* is evolved so far. However, local Deshi varieties are popularly grown and is better than Assami varieties, which are yellowish white with hard cover and gives low market return, whereas Deshi varieties are white, soft and best in eating qualities.

In a newly developed pond, the seed rate is 100-125 kg/ha whereas in old pond the seed rate is reduced to 30-40 kg/ha only since some seeds are already left in pond while harvesting the previous crops. For a successful cultivation of *Makhana*, soil type, water depth of ponds, temperature, sun shine hours, rainfall and humidity are important essential climatic factors which should be optimal to the desired levels. Approximately 20 quintals/hectare yield of raw *Makhana* is recorded and the final flake's yield is round about 8q/ha. The net income reported by the first author in his publication at the Tirhut College of Agriculture, Dholi, Muzaffarpur comes to the tune of about Rs.8,000 per hectare.

2. SINGHARA

Another popular water loving fruit crop is *Singhara* or *Paniphal*, which is called water chestnut is English and is grown throughout the world. Botanically it is known as *Trapa bispinosa* and belongs to the family *Nympheacea*. In India it is grown in Bihar, U.P., W.B., Maharastra, Rajasthan, T.N., Orissa, M.P. and J. & K. In Bihar, it is grown at large scale in many districts. *Paniphal* is eaten both without seasoning and with seasoning, and at places, its fruits are dried up for preparing its flour which is as good as wheat flour and is nutritive as well as digestive also (Table 1). Light sandy soil is not suitable for optimal yield of *Paniphal*. Soil must be heavy in texture with appreciable amount of organic matter. Newly constructed ponds are not preferred for its cultivations due to low content of organic matter in the soils. While growing *paniphal*, pond must be got clean and free from weeds. Warm and humid climate is congenial for its cultivation.

Table 1. Nutritive value of *Paniphal* fruit and flour

Ingredient		Fruit	Flour
1.	Water (%)	70	10.6
2.	Protein (%)	4.7	8.0
3.	Fat (%)	6.3	0.6
4.	Fibre (%)	0.6	1.8
5.	Starch (%)	23.3	76.4
6.	Minerals (%)	1.1	2.6

Contrary to *Makhana*, *Paniphal* may be grown in pond where fishes are cultured and thus pond utilization may be doubled. However, like *Makhana*, pond must be filled with rain water round the year and water depth must be maintained between 100 and 180 cm. Even less depth of water (60-100 cm) may be preferred provided that the depth is maintained throughout the crop period. For appreciable yield, 10t FYM and 150-200 kg ammonium sulphate per hectare may be applied to the soil before transplanting.

However, for preparing seedling, stepwise operations are followed as under :

(a) In December-January, matured fruits of *Paniphal* are stored in water to induce decay of hard fruit cover and thereafter the seeds start sprouting.

(b) The sprouted seeds are pressed inside the soils in a separate small pond having 30-60 cm shallow water. This operation is done in February-March, when the temperature starts rising.

(c) When the seedlings bear 4-5 leaves, they are used for transplanting at 1.5 m distance in 5 cm depth.

(d) In June-July, the second transplanting of seedlings is done in main pond at a distance of 2 to 3 m keeping in view the soil fertility. Sometimes 4-5 seedlings are also transplanted together.

(e) Seeds of *Paniphal* in ponds often fall down in January-February and on germination, they grow as seedlings. These seedlings may be used for transplanting in the main pond provided that the seedlings must have grown from a matured seeds to sustain the yield.

(f) In many cases, second nursery is avoided. For this, in the first nursery, seeds should not be densely raised.

(g) When the plants of *Paniphal* get fully established in the pond, 40 kg nitrogen, 40 kg phosphorus and 60 kg potash per hactare should be applied in the pond.

Plucking of fruits starts from September and continues till January-February. In general, after 20 days of flowering, fruits become matured. At the beginning, fruits are plucked at an interval of 15 days and later at an interval of 7 days. In November, plucking is normally done daily.

Like *Makhana*, no recommended variety of *Paniphal* has yet been released by the Research centres. Only local varieties are grown extensively. Important local varieties grown in Bihar are: *Kanpuri, Patania, Deshi bara, Deshi chhoti, Kotala* etc.

In general, 20-50 quintal per hactare fruits of *Paniphal* is produced. In some cases, with improved management, fruit yield of 130 quintal per hactare is also reported. This is highly profitable crop grown in wet barren land.

The main theme of the Agenda-21 and resolutions made at the 1992 United Nations Conference on Environments and Development (UNCED'S Agenda 21) focuses attention to anticipate soil, water and nutrient related problems together with other land related factors that unsustain production systems, and to intensify soil management research on well identified issues by integrating the factors involved. In the present paper, we have focussed scientific attention about a neglected area that has immense potential to produce after minimal conservation measures on survey basis. The option of growing water loving crops like *Makhana* and *Singhara* in ponds other than planting tree crops and grain crops in sloping as well as levelled land, respectively deserves detailed investigation so that a national policy may finally be framed by the Government to utilize such wastelands purposefully through growing community all over the country.

REFERENCE

Kadrekasr, S.B. 1993. *J. Indian Soc. Soil Sci. 41*, 611-622.

Singh, R.N. and Mishra, B.B. 1996. *Indian Fert. Scene Ann. 9*, 97-104.

CHAPTER - 21

DRINKING WATER QUALITY AND THE RELATED HEALTH HAZARD IN JAUNSAR, AROUND NAGTHAT IN LESSER HIMALAYA

R.B.P. Singh, Piyoosh Rautela & Dinesh Sati

En. Net Development Society, Dehradun

INTRODUCTION

Situated in the heart of the Jaunsar, Nagthat (Dabred watershed) in Dehradun district lies between 77°56' and 78°01' E longitudes and 30°34' and 30°37'N latitudes with average altitude ranging between 800 and 2200 m above mean sea level (MSL). Drinking water though not normally scarce in the region, its availability in summers is severely marred and the overall quality is questionable. Rich traditions of water harvesting together with a large number of natural springs were observed in the region. Many springs are reportedly becoming seasonal. As many as 24 seepages to the NW of Nagthat and 3 towards SW are observed. 27 spring water samples (ground water) of Dabred watershed were analysed for water quality. Most samples are acidic and can well be termed very soft. Low alkalinity with high concentration of chlorides and free CO_2 were also noticed in majority of the samples. pH, alkalinity, hardness and TSS are found to be outside the permissible limit as per ISI (1993) directives for drinking water quality. The present study aided by the Department of Science and Technology (Government of India) establishes an alarming picture of the water quality in the region and continuous consumption of this water has the potential of posing serious health hazard to the local tribal population. The observations warrants an extensive and exhaustive study to identify and manage contamination sites both from the standpoint of protecting public health and preserving the natural resources.

Jaunsar Tribal belt has not been studied in detail by environmentalists, but has always been the centre of attraction amongst the anthropologists. This region, because of its polyandry, is much romanticised than is actually understood. Situated close to Dehradun this area can be approached by motor road diverging from the Yamunotri highway. Dabred stream that drains into river Yamuna near Maror (on Yamunotri highway) forms the principal stream of this region. Small streams/*nalas* originating in the western part around Singhori, Barat khal and Chicherar are some of the major contributors to this drainage system.

As everywhere else in Uttarakhand lot of the women in this region is also miserable. They shoulder the burden of entire household chores besides the work outside the four wall of the home. Though the general health of the children in this region was observed to be good, the health of the grown ups is a cause of concern. This may be due to the addiction to the locally brewed liquor and tobacco amongst both sexes. Once a tool of adaptation, that helped the people to relax after the days hard work, liquor has taken a form of a major social evil and is a cause of concern. Therefore, the water quality data of the region warrant detailed studies because the water available to the masses is far below the normal levels.

Water ranks high amongst the priorities of any human settlement and at the same it dictates the living conditions of the people. In this region natural springs are the only cost effective and viable means of fulfilling the needs of people. At present females have to devote major portion of their time water in collection specially in the summer seasons when the pipelines fail and the supply is severely reduced. It was observed that females have to be as far as 2 kilimeters covering the rugged and steep topography to fetch water under such circumstances. Improved availability of water would lead females with ample time to devote towards other productive activities. This would improve social status of the females. Moreover enhanced availability of water can be utilised for generating energy for diverse uses.

WATER HARVESTING TRADITIONS OF DABRED WATERSHED

The water harvesting traditions of this region are magnificent and unparalleled, though not studied in detail. At the western edge of the watershed, upon a hilltop, lay the ruins of some historical monument from which the local people proudly trace their descent. The locals attribute to *Mahabharatian* king Virat, in whose court the *Pandavas* took shelter during their *agyatvaas*. Not much of interest could be found amongst these ruins except the remains of a brick lined well.

It is interesting to note that all the other structures at the hill top are made of stone while the well is lined with the fire baked bricks. In the construction of bricks very coarse sand has been used and the proportion of clay is significantly low. From its architecture the structure resembles the ones constructed for exploiting groundwater, rather than a simple rain water harvesting structure that is normally to be expected here. It was probably exploiting some herched horizon at the hill top that is made up of phyllitic rocks. The technology employed for identifying the water bearing horizon must have been highly evolved in the region. No traces of this technique could however be recollected. At present the well is defunct and has been filled up with stones.

Besides this magnificent but defunct structure, there exist a number of natural springs in the region. Most of these springs are situated in the colluvium and the state of the forest in the recharge zone is generally good.

Traditions of nature-man-sprit complex interwoven around the natural springs were observed in the region. A traditional water harvesting structure with a platform in the middle was observed in Rampur village. It is locally referred to as *pokhar or pak-ka- dhara*. The platform is associated with *Pap devata* and the water is regarded sacred. The *dhara* near Bisoi is similarly referred to as *devta-ka-dhara* and is associated with local deity Lord *Mahasu* (incarnation of Lord *Shiva*) whose wooden motif was observed at the site of the *dhara*.

Water source near Lachha is considered to be sweet and healthy by the locals. The water temperature of the *dhara* varies inversely with the atmospheric temperature. Water of Lachha-1 and *Gadiakhara* in Mundhan is believed to create stomach disorders.

More than twenty five springs were located, sampled and studied in this region as against only four depicted in the SOI maps. Most of the springs were noticed on the western side of the watershed and are reported to be of seasonal nature and dry up in the summers. The discharge of the streams was recorded in the field (Table 1). All the villages have tap water supply that often fails and the villagers have to take recourse to the springs. The water source for tap water supply is either collected from the springs or from the perennial streams. The condition of the water collection tanks was pitiable and a lot of decaying organic matter. When questioned, people of Duena village confirmed that they do not use it for drinking purposes for which the water is collected directly from the springs.

The hilly region is gifted with abundant rains but major portion of it goes waste as surface run off.

Table 1. Discharge of the springs around Nagthat

Water source, in brackets is the name of the village	Discharge (LPM)	Daily water available (LPD)
Upper Naiger (Gadaul)	7.3	10,512
Lower Naigar (Gadaul)	1.3	1,872
Teday Khadd (Gadaul)	590	849,600
Lacha-0	6	8,720
Lacha-1	43	62,525
Lacha-2	2.5	3,490
Lacha-3	2	2,280
Lacha-4	27	39,198
Lacha-5	3	4,320
Lacha-6	3	4,320
Baikhuber (Gadaul)	1.8	2,592
Hajaka (Gadaul)	29	42,027
Near Rampur	2	2,280
Pokhar/Pap-ka-dhara (Rampur)	31	45,024
Upper jogranda (Nagthat)	180	259,200
Lower Jogranda (Nagthat)	60	86,400
Duena	7	9,630
Devta-ka-dhara (Bisoi)	54	77,433
Pani-ka-dhara (Sheela)	10	14,400
Periyandu (Kuwasa)	30	43,200
Ukhniyan (Khuna Amal)	13.5	19,440
Prantkey (Khuna)	7.6	10,944
Bai-ka-pani (Kandoi)	3	4,320
Total	1114	1,603,727

As per the field as much as 1.6 MLD of water can be made available for diverse usage, to the population of the area. Besides this, there exists stream water in the area that can well be diverted for agriculture. The irony is that sufficient drinking and agriculture water is not available to the masses, and the stream water is flowing waste in almost all the areas. The productive use of the stream water must be as low as 5%, that too which are situated at convenient locations within the village.

The recharge areas of most of the streams are adequately forested, but there is ample scope for constructing chaals/khals that have the potential of enhancing the recharge of the aquifers. The water collected in the chaals/khals would be available for the grazing animal, and the effect of the same would be reflected in the lower erosional rates.

PHYSICO-CHEMICAL STUDIES

In the present investigations 27 springs and 5 sites for surface water were selected for detailed monitoring through sampling and testing for water quality and data analysis. The main objective of the study was to locate both ground water source (spring) and the surface water (stream/khal/waterfall) for the continuous assessment from potability point of view. The water samples were collected for six consequitive days between 0800 and 1200 hours and were brought to the laboratory in pyrex glass specimen bottles. Water quality parameters like pH, dissolved oxygen (DO), free carbon dioxide (FCO_2), carbonate (CO_{3-}) and bicarbonate (HCO_{3-}) were estimated at the spot immediately after collection, whereas water analysis relating to the other physico-chemical parameters were done in the laboratory of the society. The temperature was measured with mercury filled Celcius thermometer while the pH was recorded with the portable pH meter. Other parameters like DO, FCO_2, TDS, TSS, total hardness, chloride, Ca^{++} and Mg^{++} were made adopting the standard techniques (Welch 1984, APHA 1989). Correlation coefficient among the physico- chemical characteristics of both ground water and the surface water sources from different sources were also statistically correlated.

RESULTS AND DISCUSSION

Variation in physico-chemical characteristics of both surface and ground water from different locations are summarised in Tables 2 and 3. In general it was observed that both springs and the open water located at lower altitudes have different chemical characteristics in comparison to those located on the ridges. The water quality varies from place to place depending upon the characteristics of the strata and type of the underlying rocks. The time of contact of water with the bed rock may be different for different locations and this may result in some variations in the water quality of the springs (Drever, 1982, Kumar & Rawat, 1997). The details of the different water quality parameters are being discussed separately.

All values except pH are in mgL^{-1} (ppm), otherwise mentioned.

In general the water temperature closely followed the ambient temperature. In the present investigation 18.5% of the springs sample and one site of surface water showed higher value than the ambient temperature; while the range of variation between ambient and water temperature was observed higher (10°C) at the locations of lower altitude (Mehsasa village) than the springs located at higher altitudes (1.4°C). About 55.5% of the springs showed higher variation in between ambient and water temperatures.

Relatively higher variation of ambient and water temperature at Mehsasa village might be due to direct consequence of sharp solar radiation and relatively less tree cover while the lower value of the differences at the Teday Khad may be attributed to its location amongst dense forest at high altitude. Water temperature, however, has no direct effect upon human health but it governs most the chemical reactions in water. The temperature regimes regulate the kind of biota and the biological processes such as the production and decomposition in the surface and ground water.

pH

The pH of the spring waters was observed in the range of 6.3 to 7.0 (average 6.5) while in the surface waters it ranged between 5.0 to 6.6 (average 6.2). Most of the pH values for the spring water and the surface water samples exceed the permissible limit for drinking purposes (ISI 1993). In general pyrite and calcareous minerals present in the system have the greatest impact on the aqueous regime. As a result of low pH there is a danger of heavy metals dissolving easily in the water making it toxic and thus adversely affecting the drinking water quality of these sources (Singh & Choudhary, 1996).

It was interesting to record that the water from Kuasa, Rakhtar, Khuna, Chichrar and Mehsasa villages (located in lower altitudes without appreciable forest cover) have higher pH (6.7 to

7.0) values than the springs located at the higher altitudes with dense forest around. The possible reason for this could be the leaching of the organic matter (litter). Similar results were observed by Kumar and Rawat (1997) while working on the natural springs of Almora. The lowest value of pH (5.0) in the khal water may be attributed to direct surface runoff from pyritic minerals and the evaporation.

TS, TDS & TSS

Total solids (TS) is the sum of total dissolved (TDS) and suspended solids (TSS). Spring water samples were more transparent than the surface water samples as the totalsolid contents in the spring water samples were lower than the surface water samples. From the results it is obvious that the value of suspended and the dissolved solids vary considerably. The value of TDS fluctuated from 85 to 425 mgl^{-1} in spring water and from 180 to 625 mgl^{-1} in surface water samples, the suspended solids however varied between 85 to 400 mgl^{-1} and 85 to 500 mgl^{-1} respectively in spring and surface waters.

TDS in some of the water samples and TSS in the majority of the samples are found to be beyond the maximum permissible limits set by the Indian Standards (1993). Direct mixing of the percolated dissolved substances and comparatively low volume of water might be responsible for the increased concentration of the TDS while the higher values of TSS may be due to the movement of dust particles from open crop fields and the road construction sites.

Dissolved Oxygen

Concentration of oxygen in the aquatic environment is the function of biological processes such as photosynthesis and respiration and the physical processes such as water movement and the temperature. It however does not have any direct bearing on the health of the humans, but can predict the water quality of the system. The concentration of DO decreases with increase in temperature. This is explained by the increase ion biochemical reactions which consume dissolved oxygen at a rapid rate. Presence of DO is required in the waters to prevent undesirable odour (Metcalf & Eddy, 1979).

It also plays an important indicator of pollution. Low concentration of DO in the water indicates the presence of organic matter in water. In the present study DO in both spring and surface water samples was found above 7.0 mgl^{-1} in all the samples and the DO level fluctuated between 6.8 to 10.4 mgl^{-1} in surface waters and between 8.2 to 10.6 mgl^{-1} in spring waters. Water in general can thus be called fit from potability point of view, on the basis of the results obtained from DO.

Free Carbon di-oxide (FCO_2)

A considerable variation in FCO_2 concentration was observed in both surface and spring waters. It ranged between 1.0 to 6.0 mgl^{-1} and 2.0 to 33.0 mgl^{-1} in the spring and surface waters respectively. FCO_2 does not exhibit any definite trend of variation. However, the highest value of FCO_2 concentration (33.0) in the khal (surface water) is due to the combined affect of bacterial decomposition in the stagnant water and the break down of the bicarbonate into calcium carbonate and carbon dioxide. Singh and Chaudhary (1996) have also reported similar results in both ground water and surface water in the Rajmahal hills. Out of the 32 samples analysed in the present study only six samples registered higher values of FCO_2 than those prescribed by ISI (1993).

Carbonate and bi-carbonate alkalinity

In the present study carbonate remained absent but bi-carbonate was observed in appreciable amounts in both spring and surface water samples. It was observed to range between 20.0 to 95.0 mgl^{-1} and 20 to 50 mgl^{-1} in the spring and surface waters respectively. It was interesting to record higher concentration of the bicarbonate at Mundhan, Khuna, Kandoi (located at lower altitude) and at Nagthat (at comparatively higher altitude). Bicarbonate alkalinity of both surface and ground water samples in the present study was found to be well within the prescribed permissible limits (600 mgl^{-1}) of ISI (1993).

Chloride

Chlorides are readily soluble in water and are found in almost all types of waters. Washing out of salts from the soil may be a reason for the presence of chlorides in spring water. Chlorides are also reported with sulfates in water which can enter the water bodies in the form of solutions of stratified sedimentary rocks (Drever, 1982). Chloride content in the present study fluctuated in between 15.0 to 42.0 mgl^{-1} and 18.0 to 41.0 mgl^{1-} in spring and surface water respectively. Kumar and Rawat (1997) also reported

Table 2. Physico-chemical characteristics of the spring water.

Water source in brackets is the name of the village	Ambient temp.	Water temp.	pH	TDS	TSS	TS	DO	FCO_2	HCO3	CO3-	Chloride	TH	Ca++	Mg++
Upper Naigar (Gadaul)	11.2	9.4	6.3	225	90	315	7.8	1.5	-	25	21	42	10.4	3.9
Lower Naigar (Gadaul)	13.7	9.0	6.4	100	105	205	8.2	5.0	20	-	18	32	8.0	2.7
Teday Khadd (Gadaul)	13.8	12.4	6.4	150	135	185	7.0	2.5	35	-	22	40	11.2	2.9
Lacha-1	12.4	14.6	6.4	135	105	240	8.0	4.0	35	-	24	26	7.2	0.51
Lacha-2	9.8	10.6	6.4	150	110	260	7.4	5.0	30	-	23	24	6.4	1.94
Lacha-3	18.9	13.5	6.4	144	85	229	7.8	1.0	25	-	18	28	9.6	0.96
Lacha-4	18.9	14.2	6.5	185	300	485	7.0	3.0	30	-	30	32	10.4	1.45
Lacha-5	15.5	13.0	6.5	150	105	255	9.0	6.0	20	-	19	24	7.2	1.45
Baikhuber (Gadaul)	19.6	14.2	6.4	425	225	640	8.4	6.0	30	-	27	18	4.8	1.45
Hajaka (Gadaul)	19.5	13.6	6.4	400	90	490	10.4	7.0	30	-	23	28	8.8	1.45
Roagad Khadd (Rampur)	13.5	14.8	6.6	180	90	270	7.8	2.0	40	-	15	46	14.4	2.41
Pokhar/Pap-ka-dhara (Rampur)	17.5	13.0	6.5	85	95	180	7.8	2.5	40	-	24	44	14.4	1.92
Gadiakhara (Mundhan)	19.0	15.5	6.5	225	110	335	9.0	4.0	65	-	18	66	22.4	2.41
Chichrar	19.5	14.0	6.7	200	95	295	8.0	15	55	-	16	68	22.0	3.24
Upper Jogranda (Nagthat)	22.3	17.6	6.5	125	150	275	8.6	4.0	75	-	33	70	24.1	2.19
Middle Jogranda (Nagthat)	22.3	16.8	6.7	325	350	675	7.4	6.0	65	-	32	68	23.2	2.41
Lower Jogranda (Nagthat)	22.8	16.9	6.5	225	225	450	7.8	6.0	60	-	33	52	16.8	2.41
Duena	10.8	13.5	6.4	130	175	305	7.4	4.0	25	-	21	24	7.2	1.45
Devta-ka-chara (Bisoi)	14.8	15.4	6.3	85	300	385	8.2	3.0	30	-	18	24	7.2	1.45
Pani-ka-dhara (Sheela)	16.5	13.5	6.4	185	105	290	6.8	2.0	20	-	25	26	8.0	1.45
Periyandu (Kuwasa)	19.5	14.2	6.7	145	85	130	7.4	3.0	40	-	23	46	15.2	1.93

Table 2 contd...

Kulakhet (Rakhtar)	21.5	13.4	6.7	115	95	210	8.4	2.5	50	-	33	50	17.7	1.44
Timrikhet (Mehsasa)	23.5	13.5	6.8	275	85	260	7.8	5.0	45	-	22	56	15.2	4.36
Ukhniyan (Khauna Amal)	20.8	15.2	6.7	225	400	625	7.8	5.0						
Prantkey (Khuna)	20.6	15.6	6.8	325	175	500	7.4	5.0	90	-	42	120	41.7	3.85
Bai-ka-pani (Kandoi)	24.6	13.6	6.8	100	400	500	7.2	2.5	75	-	35	70	21.6	3.87
Baisgar (Kandoi)	20.2	15.1	7.0	150	375	525	7.8	6.0	95	-	37	90	31.3	2.89

All values except pH are in mgL^{-1} (ppm), otherwise mentioned.

Table 3. Physico-chemical characteristics of surface water in Nagthat

Source (near to village)	Ambient temp.	Water temp.	pH	TDS	TSS	TS	DO	FCO_2	HCO_3	CO3-	Chloride	TH	Ca++	Mg^{++}
Rain water in Khal (Lohani)	20.8	25.1	5.0	625	500	112	10.6	33.0	20	-	41.0	56	18.4	2.41
Kafed waterfall (Gadaul)	15.4	9.6	6.4	180	150	230	10.4	2.0	30	-	27.0	36	10.4	2.41
Chipli waterfall (Gadaul)	18.5	9.7	6.5	425	225	650	10.4	2.0	25	-	18.0	20	6.4	0.96
Turiyakhara nala (Rampur)	22.6	12.8	6.5	550	110	660	8.2	4.0	40	-	25.0	38	10.4	2.90
Ghasia nala (Chichrar)	23.8	14.0	6.6	560	85	645	8.8	2.0	50	-	18.0	46	15.2	1.92

higher values of chloride in the natural springs of Almora. All the samples fall within the permissible limits for chloride as prescribed by ISI (1993).

Total hardness/Ca^{++} and Mg^{++}

The hardness of water in not a pollution indicator parameter but indicates the water quality mainly in terms of Ca^{++} and Mg^{++}. Hard water is said to be one that does not form lather with soap readily. Hard water is not known to cause any disease in man (Anonymous 1980), rather protect against diseases. In the present study the values of total hardness and calcium hardness were higher than the magnesium hardness. Total hardness ranged from 18.0 to 120.0 mgl^{-1} and 20.0 to 56.0 mgl^{-1} in spring and surface water respectively.

There is no permanent hardness recorded in the present study and wide range of fluctuations were observed in both the water sources at different locations. More than 80% of the spring water and all the surface water samples were very soft to moderately soft because of value less than 50 mgl^{-1} and except for two water sources i.e. at Khuna and Kandoi villages, the rest may be termed as soft because the vaues of TH were less than 50 mgl^{-1} of $CaCO_3$ is considered very soft to moderately soft and that below 75.0 mgl^{-1} as soft, whereas TH more than 75.0 mgl^{-1} is considered moderately hard (Sawyer, 1960).

The lower value of TH may be attributed to the effect of rainfall that tends to decrease the values of bicarbonates and ultimately the hardness (Jana, 1973). Total hardness is essential from potability point of view as there is a danger of heavy metals dissolving easily in soft water with low pH, and making it toxic. All the values of TH were found to be well whithin the desirable limits as per ISI (1993) specifications for drinking water quality. Both surface and spring water is dominated by Ca^{++} relative concentration Mg^{++} concentratin ranged between 0.51 to 4.36 mgl^{-1} and 0.96 to 2.90 mgl^{-1} in spring and surface water sources respectively.

The above dominance order resembles that of the other freshwater of both Indian and foreign origin. Both calcium and magnesium in water is desirable from potability point of view. The desirable limit for clacium and magnesium are 75.0 and 30.0 mgl^{-1} as per the ISI (1993) directives. In the present study all the samples fall within this limit.

REFERENCES

APHA, 1989. Standard methods for the exmaination of water and waste water, 17th Edition, *American Public Health Association, Washington DC.*

Drever, J.I., 1982. *The geochemistry of natural waters,* Prentice Hall Inc, London.

ISI, 1993. Indian standard specification for drinking water, IS: 10500-1993, *Indian Standard Institute, New Delhii,* 6-11.

Jana, B.B. 1973. Seasonal periodicity of plankton in fresh water pond in West Bengal, *Indian Inst. Rev. Ges. Hydrobiol.*, 58 : 127-143.

Kumar, K. & Rawat, D.S. 1997. *Water management in Himalayan ecosystem - A study of natural springs of Almora,* Himvikas Publication No. 9, Indus Publishing Co., New Delhi, 151pp.

Metcalf & Eddy, 1979. *Wastewater engineering; Treatment, disposal, reuse* Tata McGraw Hill, New Delhi.

Sawyer, C.N., 1960. *Chemistry for sanitory engineers,* McGraw Hill, New York.

Singh, R.B.P. & Choudhary, S.K. 1996. Trace metal in surface and ground water in Simlong coal mining area, Rajmahal Bihar, *Proc. Nat. Sym. on Plant Diversification and Human Welfare,* 5 & 6 April 1995, Bhagalpur University.

Welch, P.S., 1984. *Limonological methods,* McGraw Hill, New York.

WHO, 1985. Guidelines for drinking water quality, *Recommendations, WHO, Geneva.*

CHAPTER - 22

INTEGRATED TECHNOLOGY MISSION FOR SUSTAINING THE PRODUCTIVITY OF SALT-AFFECTED SOILS

B.B. Mishra, I.D. Singh, J. Mall
L.N. Singh and S.P. Singh

Soil Survey and Land Use Planning Centre, Sabour,
Bhagalpur, Bihar, 813 210

INTRODUCTION

Out of an estimated area of 173.64 Mha of total degraded land in India, around 7 Mha are affected by soil salinity and alkalinity, though Yadav (1993) has put this figure as high as 26.1 Mha, in which about 3.1 Mha are occupied by coastal saline soils.

Salt affected soils are formed by natural processes including release of electrolytes by weathering, their transportation mostly by water and subsequent accumulation *in situ* or elsewhere chiefly caused by impeded drainage. High water table and faulty irrigation system resulted into widespread occurrence of the twin problem of water logging and soil salanity. Rise in water table and accumulation of *Kankar* pan shrink the normal depth of soils and accelerate the problem tremandously. Thus, among the various factors, most restricted factors which are normally uncorrectable are impeded drainage and shrinking of soil depth for facilitating water continuum with the surface layer. The present report aims at long term integrated technology mission for sustaining the productivity of salt affected soils.

Existing Technological Situation

In case of sodic (alkali) soils, the essential components of reclamation technology include proper land levelling and bunding, correct amount and recommended methodology of amendment, adequate application of irrigation and fertilizers including zinc and organic manures, correct agronomic practices and suitable crop rotation. However, in case of saline soils, efficient leaching of electrolytes (salts) and drainage to lower the water-table are essential besides judicious choice of crop varieties. Here also, provision of effective drainage is vital. However, the followings are some basic gaps to be bridged in a very friendly socioeconomic environment :

- All possible care with the scientific use of amendments like gypsum and pyrites.
- Timely follow up of management practices.
- Surface and sub-surface leaching of dissolved constituents due to impeded drainage.
- Shrinking of soil depth facilitating upwelling of dissolved electrolyte with water against vapour pressure.

Last two problems are practically uncorrectable and so pose a serious problem in effective amelioration of the salt affected soils in a big way.

Suggested Technology

Salinization is the first stage of sodiumization or sodicity or alkalinity. Obviously every possible care should be taken to remove salinization and even if sodiumization exists in soil, long term process of salt removal may lead gradually to check the sodiumization in soil, since sodium concentration in the soil solution would become lower to induce replacement of exchangeable sodium from exchange complex (Singh and Mishra, 1994).

Prevention is better than cure, so also is with correction of salt affected soils by removing electrolytes and salts on regular basis. For this, such lands should be reshaped in a systematic way as described below :

(1) Unit area of one acre should be delineated with edged boundary. By surveying, a point in the area is selected that represents maximum soil depth and which also falls in the lowest elevation so that run off may tend to accumulate therein.

(2) A pond of 20mx20mx1.5m is constructed in such a way that its boarder is no longer a ridge to facilitate runoff to accumulate in the pond. Apparently the excess soils from pond is spread over the delineated area in such a way to maintain slope towards pond. The boundary line of the pond is made covered with grasses to minimize erosion. Erosional protection of soil must be taken care of on priority basis.

(3) During rainy season, runoff containing dissolved salts falls in the pond, in which prawn can be grown successfully every year as a part of income, since prawn is tolerant to saline water. Training for maintaining ponds for prawn should be desirable among the growing community.

(4) During dry season, the pond is dry and salt through internal drainage as well as surface runoff accumulates appreciably on the surface. These salts can be scratched and dumped in a safe place. This practice should be followed periodically every year. Precaution should be given to sweep out only salts encrusted on the surface without disturbing soil. This may need wide range of experience.

(5) The salts collected and dumped safely may be screened and isolated for economic use. For example, if the salt contains Na_2CO_3 in excess, this can be used as a cheap source of soda powder (detergent) among the rural community and farmers may earn money by installing cottage industry. For this, recommended method for separating individual salts according to local requirement may be displayed among the farmers.

(6) If the salt affected unit area has *Kankar* hard pan, mechanical boring of suitable size should be done frequently within the unit to facilitate internal drainage and maintain water continuum with the surface. Through careful study, attempt should be made to use dilute HCl through boring (plastic) for facilitating dissolution of *Kankar* pan ($CaCO_3$) which may otherwise dictate the cost factor involved and the economic importance of such approach. Ultimate aim is to widen the depth of soil.

(7) The land surrounding the pond may be divided into a number of desired compartment according to slope gradient and magnitude of salinity and alkalinity. The crop cultivation and follow up crop rotation may be based on reservation of the individual compartment. Salt tolerant tree plantation may also be

preferred on the extreme upper fringe of the unit area provided that they are not interfering with normal crop cultivation and also when the salt hazard is acute.

(8) Each plot in the compartment must be reclaimed by suitable amendment like gypsum and pyrite following the recommended technology. This is further followed up with judicious application of organic matters including rice husk, green manner etc.

These integrated approaches are scientifically sound in a friendly socio-economic environment. The basic thrust is to review and examine the mission as a case study so that every aspect of technology can be specified and defined for practical utility.

Reclamation of salt affected soils is still a global problem. The integrated technology mission may successfully help to solve the problem in a big way. This mission is a long term approach and Government should encourage the mission by formulating policy. The mission needs verification and specification so that approved technology may be displayed among farmers for follow up.

REFERENCES

Singh, V.N. and B.B. Mishra, 1994. Sodiumization of some Alfisols in toposequence occurring in Indo-Gangetic plain of Bihar, *J. Indian Soc. Soil Sci.* 42 : 626-633.

Yadav, J.S.P. 1993. Salt affected soils and their management with special reference to Uttar Pradesh, *J. Indian Soc. Soil Sci.* 41 : 623-629.

CHAPTER - 23

CEMENT INDUSTRY AND ITS IMPACT ON AGROECOSYSTEM

D.D. Pandey & Sanjeev Kumar

Department of Botany

Nalanda College, Biharsharif - 803 101

INTRODUCTION

Enormous amount of cement dust is deposited on the leaves of agroecosystem in the vicinity of the Cement Factory. The particulates going into the atmosphere, depending on their size and weight, may remain in air for varying length of time. Those larger than 10 in size settle under forces of gravity on surface of vegetation and soil but the small one remains suspended in air for longer periods of time and in accordance with the gas laws gets dispersed, distributed and diffused by wind motion and eddy currents. The suspended particulate matter (SPM) gradually gathers mass through agglomeration, coalescence and water vapour deposite and eventually settles down on surface or may be washed down in rain. In comparison to gaseous air pollutants, many of which the cause of injury to various types of vegetation, relatively little is known about and limited studies have been carried out on the effect of particulate air pollutants on vegetation (Rao, 1971; Periasamy and Vivekanandan, 1982; Singh, 1983; Kunwar, 1984; Agrawal *et al.*, 1988; Pandey and Rastogi, 1992; Tiwari and Chaudhury, 1993; Chauhan, 1994; Singh *et al.*, 1994; Mehta and Mahto, 1994; Pandey and Satya Nand, 1995. Cement dust is a serious air pollutant in the vicinity of agricultural crops of A.C.C. Cement Factory, Jhinkpani, Chaibasa, Bihar. The chemical nature of toxicity of agroecosystem from cement dust pollution has not been studied in detail as yet. The present study deals with the effect of cement dust pollution on biomass, chlorophyll and grain characteristics of rice.

MATERIALS AND METHODS

Rice is grown by the villagers in the vicinity of cement factory Jhinkpani, Chaibasa, Bihar. Dust load was estimated at different distances (1 Km to 5 Km) away from emission source in the prevailing wind direction (SW --- NE) by dust collecting jar method (Rao, 1985).

Rice seeds were sown in an area of 50m x 50m at a distance of 100 m in the north east direction of cement factory in the middle of July, 1995. The second plot was selected away from cement factory where dust load was zero and treated as the control one. The samples were collected from plots of control and polluted zone at the age of 15, 30, 45, 60, 75, 90, 105, 120 and 135 days. Biomass was estimated by the method of Misra (1968). Concentration of chlorophyll-a and b in leaves of rice at different ages of control and polluted were determined using the formula of Machlachlan and Zalik (1963). The grain characteristics were analysed through the method as given by Misra (1968).

RESULTS AND DISCUSSIONS

The biomass of rice plants at any stage of growth of cement dust polluted plants was lesser than that of control one (Table 1). The maximum biomass was recorded at the age of 120 days i.e. 49.30 g/plant (control) and 42.79 g/plant (polluted). The cement dust pollution had reduced 13.20 per cent biomass indicated a reduction of photosynthesis. This reduction in biological yield of polluted plants could be attributed to the cement dust which possibly affect the metabolism of plants in several ways. Thus, several extra and intercellular changes in leaf may cause serious metabolic interference in polluted plants leading to reduction in biomass (Agrawal *et al.*, 1988).

The concentration of chlorophyll a, b and total chlorophyll in unit weight of polluted leaves were always lower than those in the controlled leaves (Table 2). When compared with the control, a maximum reduction of 21.80 per cent of total chlorophyll was recorded in 90 days old plants. The concentration of chlorophyll b also continued to increase in both controlled and polluted plants till the age of 90 days. The chloroplast damage by incorporation of cement dust solution into leaf caused reduction in the chlorophyll concentration. Similar results were obtained with cement dust in maize leaves (Pandey and Simba, 1990).

Table 1. Standing crop biomass (g/plant) of Control (C) and cement dust polluted (P) rice plants at various stages of growth period.

Days	Shoot (s)											
	Stem		Leaf		Inflorescence		Total		Root (r)		Total (s+r)	
	C	P	C	P	C	P	C	P	C	P	C	P
	0.58	0.48	1.48	1.19	-	-	2.06	1.67	0.41	0.36	2.47	2.03
15	±0.05	±0.02	±0.09	±0.06			±0.20	±0.08	±0.02	±0.01	±0.24	±0.21
	0.99	0.84	1.86	1.51	-	-	2.85	2.35	0.73	0.59	3.58	2.94
30	±0.08	±0.05	±0.14	±0.12			±0.22	±0.20	±0.06	±0.04	±0.29	±0.22
45	3.63	2.91	3.38	2.77	-	-	7.01	5.68	2.08	1.68	9.09	7.36
	±0.33	±0.22	±0.31	±0.29			±0.61	±0.47	±0.20	±0.10	±0.82	±0.60
60	5.51	4.46	4.57	3.74	-	-	10.08	8.20	2.19	1.83	12.27	10.03
	±0.50	±0.34	±0.39	±0.38			±0.82	±0.78	±0.29	±0.09	±1.22	±0.93
75	6.55	5.38	7.83	6.39	-	-	14.38	11.77	3.35	2.88	17.73	14.65
	±0.53	±0.46	±0.72	±0.52			±1.47	±1.30	±0.30	±0.23	±1.81	±1.64
90	18.13	13.05	10.59	8.67	3.31	2.81	32.03	24.53	3.69	3.11	35.72	27.64
	±1.52	±1.02	±1.05	±0.68	±0.23	±0.22	±3.81	±2.16	±0.31	±0.24	±3.22	±2.42
105	21.52	14.66	11.18	8.88	8.85	7.47	41.55	31.01	3.77	3.24	45.32	34.25
	±1.90	±1.30	±1.06	±0.87	±0.77	±0.68	±4.04	±3.30	±0.32	±0.33	±3.84	±2.70
120	19.01	17.98	11.17	9.15	12.61	10.37	42.79	37.50	6.51	5.29	49.30	42.79
	±1.61	±1.62	±1.03	±0.91	±1.10	±0.81	±4.00	±3.32	±0.62	±0.58	±4.22	±4.01
135	17.77	15.93	10.29	8.43	14.65	12.44	42.71	36.90	5.92	4.96	48.63	41.86
	±1.70	±1.56	±0.89	±0.70	±1.42	±1.10	±4.09	±4.12	±0.50	±0.42	±5.59	±4.84

There were considerable changes in the characteristics of grains obtained by cement dust polluted plants. The number of grains per spike was reduced resulting a loss of 12.08 per cent in the yield. The weight and volume of 1000 grains were reduced 11.12 and 8.94 per cent respectively (Table 3). The reduction in grain yield was due to the failure of pollen grain germination on the dust laden stigma and fertilization (Rao, 1971; Pandey and Rastogi, 1992). The grains from polluted plants and reduced amount of protein may be due to reduced photosynthetic potential of polluted plants (Pandey and Satya Nand, 1995). It may be concluded that rice cultivation could suffer a loss in size, in terms of both quantity and quality in cement dust polluted area.

Table 2. Chlorophyll contents (mg/g fresh leaf) of control (C) and cement dust polluted (P) rice plants at different stages of growth.

Age (Days)	Site	Chlorophyll a	Chlorophyll b	Total chlorophyll
15	C	1.32	0.72	2.04
	P	1.29	0.68	1.97
30	C	1.45	0.79	2.24
	P	1.34	0.71	2.05
45	C	2.17	0.87	3.04
	P	1.80	0.75	2.55
60	C	3.50	2.40	5.90
	P	3.00	1.61	4.61
75	C	4.09	3.49	7.58
	P	3.37	2.09	5.46
90	C	4.68	3.67	8.35
	P	3.71	2.82	6.53
105	C	4.40	3.54	7.94
	P	3.50	2.59	6.09
120	C	3.49	2.99	6.48
	P	3.10	2.32	5.44
135	C	1.55	1.49	3.04
	P	1.11	0.75	1.86

Table 3. Characteristics of grains obtained from control and cement dust polluted rice plants.

Characteristics	Control	Polluted
Number of grains/spike	1496.23	1315.11
Weight of 1000 grains (g)	18.440.74	16.390.65
Volume of 1000 grains (cc)	19.670.79	17.910.88
Moisture (%)	12.770.44	11.020.41
Protein (%)	15.590.50	14.110.53

REFERENCES

Agrawal, S.B., M. Agrawal, and P.K. Nandi, 1988. *Indian J. Environ. Hlth.* 30(4) : 340-347.

Chauhan, U.K. 1994. Proc. Acad. Environ. Bio. 3(2) : 207-211.

Kundaw, J. 1984. Particulate matter. *Scavanger* 1(1&2) : 11-14.

Machlachlan, S. and S. Zalik, 1963. *Can. J. Bot.* 41 : 1053-1062.

Mehta, U.C. and M. Mahto, 1994. *Polln. Res.* 13(1) : 89-90.

Misra, R. 1968. *Ecology Work Book.* In : Oxford and IBH Publishing Co., New Delhi, 244.

Pandey, D.D. and S. Rastogi, 1992. *Environ. & Ecology* 10(4) : 980-982.

Pandey, D.D. and Satya Nand, 1995. *Environment & Ecology* 13(4) : 901-903.

Pandey, D.D. and A.K. Simba, 1990. *Geobios.* 17 : 40-41.

Periasamy, K. and M. Vivekanandan, 1982. *Comp. Physiol. Ecol.* 7 : 157-160.

Rao, D.N. 1971. Academic Press. In : Proceeding of 2nd *Int. Clean Air Cong.* Academic Press. 273-276.

Rao, D.N. 1985. In G.V. Subrahmanyam *et al.* Editor *Air Pollution* and *Plants*, New Delhi, India, 10-45.

Singh, C., H.K., Parwana, R.S. Singh, and S.S. Marwaha, 1994. *Pollen Res.* 13(1) : 79-81.

Singh, S.N. 1983. *Environ. Pollut.* 32 : 171-177.

Tiwari, T.N. and S. Chaudhury, 1993. *Indian J. Environ. Prot.* 13(11) : 821-824.

CHAPTER - 24

OPEN CAST COAL MINING IN DHANBAD : EFFECTS ON CROP PRODUCTION

R. Prasad & B.K. Roy

Department of Botany, Banaras Hindu University, Varanasi-221005, India

INTRODUCTION

The mining industry is an important point source of metals in the environment, causing chemical pollution and results in the localised disturbance of large areas of land. Numerous investigators and observers have made a comment that the destruction of soil properties might also significantly alter the hydrologic system and deplete the nutrients from mining system (O'Neill *et al.,* 1977; Parkinson, 1979). Generally, the effects of soil disturbance are more adverse than beneficial, because useful soil has been developed by soil forming processes that require hundred to ten thousand years to reach a steady state resulting in maximum productivity (Schaffer, 1984). On the other hand, Omodt *et al.,* (1975) noted that if similar ranges of pre mining soil properties are replaced on mine land, successful reclamation is neither impossible nor as difficult but the ability of achieving revegetation success is mainly depend on mine soil quality. Therefore, industry research bad regulating personnels have taken the task and sometimes compared physical and chemical properties of mine spoils with base line native soil characteristics, and considered as one of the method to judge reclamation success. Indorante *et al.*, (1981) have compared mine- land soil properties of five reclaimed sites in Illinois and three unmined sites. U.S.A. regulatory guidelines of the rehabilitation of the surface mined land was made, which required the assessment of soil chemical qualities both before and after mining (Imes and Wali, 1978). Assessments were mainly based on (a) characterisation of pre-mining land scape so that the magnitude of any changes can be evaluated (b) determination of the suitability for plant growth on mine spoils and (c) modification of the related practices to achieve better rehabiliation results.

From the agricultural point of view, earlier few results have been achieved on the basis of the study of major soil forming processes correlating with mine soil formation, development and the production potential of these soil in UK (Kimber *et al.*, 1978). Toy and Shay (1987) have suggested that physical properties like soil moisture, bulk density, organic carbon and nutrients availability are major factors in governing the establishment of plants. The first practical step in the process of reclamation, disturbed mobilisation and availability of elements and their subsequent uptake by rehabilitation species on mine spoils have been usually considered. Severson and Gough (1983) and others have analysed and compared the chemistry of plants and soil from on and off mine sites and showed the essentials to identify the classified possible variations, if later stages of reclamation programs are to succeed.

The reclamation of mine spoil, for agricultural or recreational purposes is now common in coal mining areas of Britain. Initial attempts to reclaim coal mines treated to be empirical and followed established agricultural practices (Coatis, 1956). Later on extensive researches have been carried out on the factors limiting plant growth (Goodman and Barry, 1975) and accordingly other more systematic approaches have been adopted (Down, 1975; Kimber *et al.*, 1978). Cornwell (1973) used saturation extracts to measure the nutrient status in spoil and supported these analysis with bioassay and plant growth tests. Fitter *et al* (1974) showed the variability in spoil constituents leading to differences in plant growth and yield.

Therefore, in the present study, investigation has been carried out to understand the quality of mined spoil influenced by open cast mining activities. Distribution patterns of essential and non-essential elements and nutrients in the studied region were demonstrated by comparing the mine spoil to control soil. In field study, the impact of coal mine spoils on *Cajanus cajan* with reference to growth responses and standing crop yield were also determined.

MATERIALS AND METHODS

Study area

The study area is located at 23°30'N-23°55'N in Eastern and Western coal field area of Dhanbad, India. The climate is tropical and influenced by monsoon conditions. The minimum and maximum temperature is 6C-18°C and 39°C. The average rainfall is 1072mm (according to meteorological data).

Distance between these two selected areas is about 53 km. From Western coal field, two important open cast mine sites i.e. Modidih and Kusunda have been selected, which are about 20-30 years old and estimated production of coal is about 1.5-1.9 M tons per annum. Another two important coal mine sites selected from Eastern coal field area are Mandman and Rajpura, which are about 20-25 years old and estimated coal production is about 1.0- 1.5 M tons per annum. Control site (Topchanchi), an agricultural area is located at 40 Km away from coal mines.

Mine sites were selected on the basis of following criteria : (1) vegetation started since last 5-8 years old, (2) Present spoils are being used in rehabilitation practices, (3) grasses and legumes are dominant vegetation.

Soil sampling

Spoil samples were collected from 0-30 cm depth, randomly from 10 locations at 10 meters interval from mine and control sites. Spoils were attached to the roots of the plants other than grasses. Samples of each site were mixed separately and large pieces of spoil were discarded through a stainless steel screen with 1cm opening. Further all samples were transported to the laboratory for analyses.

Soil analyses

Mine spoil and control soil samples were dried at ambient temperature and then disaggregated in a mortar and pestle and passed through 2mm sieve. Following analytical methods were used for the determination of constituents present in spoils.

Soil pH was determined with a glass electrode by preparing 1:1 mine spoil and water mixture according to the method used by Peech (1965). Available P in spoil was estimated by extraction in $NaHCO_3$ (pH 8.5) and determined by using ammonium molybdate stannous chloride method (Sparling *et al.*, 1985). NO_3-N was determined by phenol disulphonic acid method (Jackson, 1958). Available potassium was determined by 'mediflame 127' flame photometer (Jackson, 1958). Organic carbon was determined after digesting in $K_2Cr_2O_7$ and H_2SO_4 and by using Walkley and Black's rapid titration method (Jackson, 1958). Moisture in soil samples from each location was determined in triplicate. About 25gm fresh samples were taken in a beaker, dried in an oven for overnight at 105°C and after weighing moisture content was determined and calculated in per cent (USDA Hand Book, 1977).

For heavy metal analysis, about 1.0 gm of each dried soil samples were placed in a beaker and added with total volume 7.0 ml of Tri acid (HNO_3 : H_2SO_4 : $HClO_3$, 5 : 1 : 1) mixture and left for digestion for 12 h. After complete digestion, samples were diluted with 50 ml double distilled water and filtered. Concentration of metals determined with the help of Atomic Absorption Spectrophotometer (Perkin Elmer model 2380) (Jackson, 1958) and expressed in g/g dry wt.

Determination of plant biomass and yields

Five study locations were selected at four mine sites and single control site. Selected mine spoils were rolled and labeled and study plots of 8' x 8' size were made. Seeds of *C. cajan* (var. Pusa Bahar) were collected from Institute of Agricultural Sciences, B.H.U. Varanasi. and sown in 16 plots of mine sites (four at each sites) and 4 plots of control sites in three replicate. Planting dates and other management practices were variable, however, these variables were constant for all reference sites. An individual field under the same farm management practices with each mine such variables, as hybrids, planting dates, soil fertility, soil type and other management factors. Seeds were shown in July and harvested in month of March of each year. Irrigation was performed according to requirements.

For plant biomass, plants were harvested from ten random meter- square quadrates within each plots from all mines by up-rooting plants from each quadrate at 30, 90, 150 and 210 days growth periods. Samples were washed with distilled water and dried at 80°C in hot air oven for oven night and total biomass was determined and expressed in g dry weight.

Yield of *C. cajan* was measured by hand harvesting two adjacent rows at sampling sites selected to give representative yield for mine spoils. Yield was based on total number of pods, number of seeds, yields/plant and weight of 100 seeds and corrected to 15% moisture. Differences between yield for control and mine areas at a site served as an estimate of the yield reduction for spoils assigned at the each site. Yield reduction for each spoil class were calculated using the following formula :

Percent yield reduction = 100
(Control yield - Spoil yield)/control yield.

The yield reduction in terms of each spoil were averaged to give estimate for individuals. A total 40 samples of seeds in 1995, 60 in 1996 and 80 in 1997 were harvested for estimation of yield.

Statistical analysis

Student's 't' test was used to test the significant difference in physico-chemical constituents of mine spoils from control soil. One way ANOVA was applied to test the significant difference in nutrients and heavy metal contents among spoils of all mine sites. Correlation coefficient @ was used to observe the relationship among all the constituents in mine spoils of all sites. Two way ANOVA was also used to test the significant differences in yield between sites and years.

Results

Table-1 summarizes the physico-chemical properties of coal mine spoils and control soil. The texture of mine spoils was sandy loam with 2.0, 56.9 and 41.1% gravel, sand and clay respectively, whereas control soil had 2.4, 72.3 and 25.3% gravel, sand and clay respectively. A little differences in gravel content occurred between control and mine soil but spoils of all mine sites had higher content of clay than control soil. Mean pH varied from 5.0 to 5.6 and showed significant ($P > 0.05$) differences than control (pH 6.5). Moisture content was 15.43% in control soil, whereas it varied from 7.46 to 8.85% in mine spoils and showed significantly lower than control soil. Organic carbon was recorded 0.46% in control soil, whereas it varied from 0.22 to 0.31% among all mine spoils and showed significant (P) decrease than control soil. In control soil, available phosphorous was maximum (2.5 ppm), and it was significantly less in mine spoils than control soil (P). Available nitrogen varied from 17.4 to 26.4 ppm in mine spoils, whereas, it was highest (41.2 ppm) in control soil, and showed significant decrease than control soil (P). Similarly, available potassium was highest (45.4 ppm) in control soil, whereas it varied from 34.4 to 41.6 ppm in all mine spoils and showed significant decrease than control soil (P). All the spoils had the lowest pH, moisture content, organic carbon, available P, K and N at Kusunda site (in Western coal field), and highest at Rajpura site (in Eastern coal field).

Table-1. Physico-chemical properties of coal mine spoils and control soil.

Sites	pH	Texture	Moisture content (%)	Organic carbon (%)	Available phosphorous (ppm)	Available nitrogen (ppm)	Available potassium (ppm)
Control	6.50	Sandy	15.43	0.64	2.50	41.20	45.40
	.071	loam	0.031	0.016	0.13	1.32	0.51
Modidih	5.20	Sandy	7.60	0.26	1.70[b]	18.80	37.20
	0.066	loam	0.049	0.017	0.11	0.66	0.86
Kusunda	5.00[d]	Sandy	7.46	0.22	1.60[c]	17.40	35.40[d]
	0.080	loam	0.033	0.013	0.08	0.81	0.75
Mandman	5.30[d]	Sandy	8.05	0.27	1.80[b]	21.00[d]	40.80[c]
	0.084	loam	0.036	0.014	0.89	0.74	
Rajpura	5.60[d]	Sandy	8.85[d]	0.31	1.90[b]	26.40[c]	41.60[b]
	0.051	loam	0.033	0.09	1.20	0.93	

Mean values of five replicates SE.

Data marked are significantly differed from control according to student's 't' test (a not significant; b, P; c, P; d, P).

ANOVA showed the distribution pattern of nutrients, organic carbon and heavy metals in spoils (Table 2). There was no homogenity of variance between spoils, as the composition have differed, may be due to aging or checking of spoil erosion during vegetation. Significant differences among the samples of spoils were observed. Spoil pH was found with F-ratio 10.16 (P); moisture content, F-ratio 63.29 (P); organic carbon, F-ratio 7.65 (P); Nitrogen, F-ratio 1861 (P); Phosphorus, F- ratio 2.9 (P) and Potassium, F-ratio 12.18 (P). Statistical analysis also indicated the significant variations in heavy metal contents of Cr^{2+}, Mn^{2+} and Fe^{2+} among mine spoils. Among them only available P^{2+} and Cd^{2+} showed non-significant variation.

Table-2. Analysis of variance between all constituents levelc in coal mine spoils.

Properties	F-Ratio	Prob. level.
pH	10.16	0.0005
Soil moisture	63.29	0.00
Organic carbon	7.65	0.0022
Available N	18.61	0.00
Available P	2.902	0.067
Avialable K	12.815	0.0002
Cromium	2.373	0.108
Nickel	72.256	0.00
Cadmium	1.239	0.328
Zinc	54.28	0.00
Copper	8.977	0.001
Lead	12.26	0.0002
Manganese	533.86	0.00
Iron	3.32	0.0468

Correlation matrix was applied to observe the relationship among all variables of mine spoils (Table 3). Spoil pH showed positive correlation with organic carbon (P), soil moisture, available N, P and K but negative correlation with all metals. Spoil moisture showed positive correlation with organic carbon, available N, P and K. Similarly, organic carbon showed positive correlation with available N (P), P (P) and negative correlation with all metals. Available N, P and K showed negative correlation with all metals, but a positive correlation was observed between P and K (P). Statistical analysis was used in order to establish the process of heavy metals were positively correlated with each other, except Cr^{2+}, Cu^{2+} and Pb^{2+} which were negatively correlated with Fe^{2+}. Mn^{2+} was present in high amount in spoils of all mine sites and had positive correlation with all metals. Zn^{2+} also retained the similar relationship with all metals as Mn^{2+}. Concentration of Fe^{2+} was also maximum to comparison to other metals and showed negative correlation with Cr^{2+}, Cu^{2+} and Pb^{2+}.

Based on the statistical analyses, it was concluded that there was significant variation in nutrients contents and metal concentrations among all mine sites except Cd^{2+} and Cr^{2+} (Table 3). Above all results conclude that metal levels were higher in mine spoils and found significantly increased than control soil (P, P) and studied sites can be arranged in order to their metal level, Kusunda Modidih Mandman Rajpura control.

Plant biomass

Table 4 summarises the per cent reduction in biomass production of *C. cajan*, grown in coal mine spoils at 30, 90, 150 and 210 days growth. Mean biomass production was markedly higher in control soil than mine spoils. Maximum reduction (41%) at Kusunda spoil and minimum (25%) at Rajpura spoil was observed in comparison to control soil at 210 days growth period, while inhibition had initiated with 25% and 19% decrease at both sites on 30 days respectively. Plant biomass showed significant decrease (P, P) at all mine sites in respect to control site. Result showed that the effect of mine spoils on biomass production was related to the constituents level present in spoils and growth periods.

Table 4. Pre cent reduction in plant biomass of *C. cajan*, grown in coal mine spoils at different sites at 30, 90, 150 and 210 days growth period.

Sites	30 days	90 days	150 days	210 days
Kusunda	25^{c}	30c	39d	41^{d}
Modidih	21^{b}	24c	30c	31d
Mandman	20^{b}	23c	28c	35d
Rajpura	19^{b}	20b	24c	25c

Data marked are significantly differed from control using student's 't' test (a = not significant, b = $P < 0.05$, c = $P < 0.001$, d = $P < 0.001$).

Yields

Table 5 shows the yield production of *C. cajan* exposed to mine spoils. Yield reduction was significant at all mine spoils than control soil (P). Yield reduction found at individual spoils marked the types of mining. In 1995-97, yield reduction was high at all the mine sites than control sites. Maximum yield reduction occurred at Kusunda site indicated the severity of mine spoils and grain yields

Table-3. Correlation coefficient among constituents level in coal mine spoils.

Properties	pH	SM	OC	AN	AP	AK	Cr	Ni	Cd	Zn	Cu	Pb	Mn	Fe
pH	1.0000													
SM	0.6614	1.0000												
OC	0.9991***	0.6848	1.0000											
AN	0.9743	0.7897	0.9766**	1.0000										
AP	0.9496	0.7587	0.9614	0.9291	1.0000									
AK	0.9252	0.7124	0.9382	0.8866	0.9944**	1.0000								
Cr	-0.5492	0.2517	-0.5269	-0.3578	-0.4208	-0.4557	1.0000							
NI	-0.8223	-0.1592	-0.8123	-0.6743	-0.7555	-0.7791	0.9122	1.0000						
Cd	-0.9340	-0.3504	-0.9224	-0.8384	-0.8308	-0.8245	0.8090	0.9571*	1.0000					
Zn	-0.6974	-0.1776	-0.6682	-0.6761	-0.4378	-0.3784	0.5948	0.6195	0.7723	1.0000				
Cu	-0.7134	0.0081	-0.7000	-0.5383	-0.6404	-0.6758	0.9626*	0.9851**	0.9009	0.5655	1.0000			
Pb	-0.5432	-0.0135	-0.5469	-0.3594	-0.6183	-0.6900	0.7897	0.8547	0.6877	0.1260	0.8831	1.0000		
Mn	-0.8019	-0.1223	-0.7904	-0.6476	-0.7311	-0.7567	0.9264	0.9993***	0.9485*	0.6146	0.9907	0.8597	1.0000	
Fe	-0.5078	-0.5958	-0.4971	-0.6523	-0.3371	-0.2361	-0.1327	0.0476	0.3327	0.7032	-0.0850	-0.4374	0.0238	1.0000

SM = Soil moisture, OC = Organic carbon, AN = Available nitrogen, AP = Available phosphorus, AK = Available posatassium.,

, * Significant at the 0.05, 0.01 and 0.001 levels respectively.

Table-5. Spoil-induced percent reduction in yield of *C. cajan* at different coal mine sites in 1995, 1996 and 1997.

Site	No. of pods/plant				No. of seeds/plant				Yield/plant				Wt. of seeds (g)			
	1995	1996	1997	Ave.	1995	1996	1997	Ave.	1995	1996	1997	Ave.	1995	1996	1997	Ave.
Kusunda	52	47	44	47.8^{d}	69	65	62	65.3	53	48	45.5	48.8	49	44.5	42	45.2
Modidih	42	38	35.5	38.5^{d}	63	59.5	57	59.8	51	47	44.5	47.5^{d}	45	41	38.5^{c}	41.5
mandman	40	36	33.5	36.5^{d}	62	59	57	59.3	49	45	43	45.6^{d}	40	36	34	36.6^{d}
Rajpura	24	21	19	21.3^{c}	50	48	46.5	48.2	34	31	29	31.3^{c}	28	26	24.5	26.2^{c}

Data marked are significantly differed from control using student's 't' test (a = not significant, b = P0.05, c = P0.01, d = P0.001).

were expressed by spoil effects averaged for each of 3 years. The weighted average yield reduction in *C. cajan* at sites for 3 years were 48.8%, 47.5%, 45.6% and 31.3% for Kusunda, Modidih, Mandman and Rajpura respectively. These results include both the effects of relative extent of spoil-affected site directly over coal mines and magnitude of yield reductions for each mine spoil effect for each year. Yield reduction exhibited by spoil types effects interaction. Large reduction occurred on severe spoils and was 1.6 times greater for severe class th· n moderate class.

DISCUSSION

Presently, surface coal mining area concerns over the ability to achieve revegetation success that frequently governs by mine soil quality. Earlier, comparison of physical and chemical properties of mine spoils with native soil was made to adjudge the reclamation success in India (Banerjee, 1989). Distribution pattern of different metals in Indian coal mines spoils have been reported by Rath *et al.*, (1994). For agricultural purposes, coal mine spoils have been characterised with low pH, associated with toxic concentration of certain metals, low nutrient status (N, P, K and organic carbon), high surface temperature and moisture stress in different other countries (Singer and Stumm, 1970; Gough and Severson, 1983; Richardson and Greenwood, 1973).

In the present study, the texture of coal mine spoils and control soil is sandy loam but differences are in content of rocky stone only. The sand content in control soil was 35% higher than the mine spoils. Conversely, the spoils on slopes exhibited 7-8% silt and clay higher than control soil as in case of other coal mine spoils composition (Toy and Shay, 1987; Banerjee, 1989). These changes indicate a positive indication of revegetation of selected mines surfaces. The water holding capacity in mine spoils was found lower than control soil, however established grasses and herbs vegetation on spoils might increase the water holding capacity. Soil moisture present in mine spoils may act as a limiting factor for the improvement of water holding capacity during vegetation.

Variations in pH from 5.0 to 5.6 in mine spoils, which is less than control (6.5) and supported as in case of other coal mine spoils (Berg and Vogel, 1973; Down, 1975), are under limit for plant growth. However, mean pH of spoil is about 0.9 to 1.5 units below than the mean pH of control soil and none of spoils have shown pH values less than 4.0, therefore, the above pH values might be considered more acidic than expected in the environment.

In the present study, impact of mine spoils on reduction in biomass and yield production in *C. cajan* are mainly due to following soil factors : (i) lower pH, (ii) very high iron levels, (iii) extremely low water holding capacity, (iv) deficiency of organic matter and other nutrients (N, P, K), and (v) high concentration of elements in spoil. Level of elements (essential and non-essential) in mine spoils were high with about 2-fold increased than control soil, and Mn and Fe had maximum levels. Although these factors are not satisfactorily interrelated, because they may well have common origin in weathering reactions of pyrite. Pyrite oxidation and hydrolysis give rise to large amounts of H^+ ions and by decomposition exchange reactions with other spoil minerals results to increase concentration of Fe, Mn, Zn, Cu and Ni in spoils (Barnhisel and Rotromel, 1974; Elliot *et al.,* 1986). Among them, high Fe concentration are possibly not the results of breakdown of pyrite but may be due to decomposition of carbonate minerals. Siderite and ankerite which are often found in spoil (Sach *et al.,* 1975). High Mn concentrations are also in part due to the decomposition of ankerite. Concentration of organic carbon and other nutrients are low, may be due to formation of other insoluble complexes between meatls and organic matter upon decomposition because they have both cation exchange capacity and chelating ability to form complexes (Chen and Stevenson, 1986). Findings of distribution and the relative deposition of metals and their correlation among spoils are possibly due to changing time and geological environment at each location (Gough and Severson, 1981). In this study, means and deviation for each constituent in spoils of mine sites might be a useful designed guidelines during rehabilitation program because all the constituents level are under the toxic concentration value (Bowen, 1978). Estimation of wide spread deficiencies of major nutrients, particularly N, P, K concentration and organic carbon are very low but similar to the findings of agro-based in mine spoils (Fitter *et al.,* 1974; Withams, 1975; Chadwick, 1973). These factors obviously directed to poorest growth and yielding as the most limiting factors and responded as similar as in case of other vegetated mines (Tasker and Chadwick, 1978). For easier and detail observations, samples from all sites have been categorized into two homogenous groups on the basis of findings in spoil-induced yield reduction and biomass production. Highly reduced biomass production and yield at sampling mines area have low pH, high metal concentrations and less nutrients, categorised as severe class. Whereas, other sampling sites show comparatively less inhibition, having less element concentrations and increased nutrients and pH than severe class, categorised as moderate class. This means that growth and yields reduction found in each individual spoil effect class were dependent on variability in soil composition, or other several factors. In this respect it requires to identify the micronutrients to crop responses. In present result, growth inhibition in plants varied with spoil type at the availability of elements and organic matter and in this respect metal absorption by promoting diffusion of metals from soil to root in severe sites has been described (Bell and Trimura, 1991; Lindsay, 1974). At the same time low pH and texture have also affected the metal conviction and diffusion of metals during the plant growth (Kabata-Pendias and Pendias, 1994).

In 1995-97, observed yield reduction of *C. cajan* is comparatively higher in all mine spoils but gradually improved and varied with the individual years. We found over all 5-8% improvement of average yield than over previous year reduction in affected yield. This relative impact correlate the possiblity of metal precipitation, changing in mining activity, weather and crop sample areas. In future, these condition may lead to natural process of chemical changes in spoils with the help of factors like rapid addition of biological components such as root exudates and microbiological activities. At present, young stage of spoil development is indicated and therefore, changing in physical and chemical properties should be expected and that may even exceed to level of control soil in the long term. Sampling areas were physically uniform in terms of surface instability, erosion, rainfall and incoming solar radiation but other physical factors such as particle size, water retention capacity and compaction were more variable. These factors can also exert large influence on growth and yield (Ayerst, 1977). Temperature and water stress are severe in summer and autumn, and spoil pH is also low in comparison to other seasons.

In present work, sampling was carried out in winter and therefore, none of the values recorded should be regarded as the most extreme encountered in any year. Dominant elements in spoils seem to be the fractions of importance in assessing the micronutrient supplying capacity of mine spoils. Where as soil pH, organic and nutrient level alter the distribution of elements in the plant available pools (Sims, 1986). Differences between growth and yields on metal contaminated spoil also itself explains the multiple metal tolerance mechanism in this species at its maturity or high degree of tolerance may be a species characteristics. (Baker, 1978, Cox and Hutchinson, 1980).

It thus can be concluded that *C. cajan* can be a suitable species to vegetate on selected mine spoils on the basis of the result of growth and yields performance and finally this species can be considered from reclamation point of view.

ACKNOWLEDGEMENTS

The authors are greateful to Ministry of Environment and Forests, Government of India, New Delhi for financial assistance in research and preparing the manuscript, and Director, Central Mining Research Institute, Dhanbad, for his help in collecting the samples.

REFERENCES

Ayerst, J.M. 1977. The effect of compaction of coal shale on the revegetation of spoil heaps. Reclamation Review (Journal of Canadian Land Reclamation Association) in Press.

Baker, D.E. 1974. Copper : Soil, Water, plant relationship. Fed. Proc. 33 : 1188-1193.

Baker, A.J.M. 1981. Accumulators and excluders-strategies in the response of plants to heavy metals. Jn. Plant Nutr. 3 : 643- 654.

Banerjee, A.N. 1989. Reclamation and land laid waste by mining. Wastelands New 4(3) : 13-22.

Barnhisel, R.I. and Rotromel, A.L. 1974. Weathering of clay minerals by stimulated acid coal spoil-bank solutions. Soil Science, 118 : 22-7.

Bell, R. and A.H. Tremura 1991. Soil Metal effects on the germination and survival of *Quereus alba* L. and *Q. prinus* L., Environ. Exp. Bot. 31 : 145-152.

Berg, W.A. and W.G. Vogel 1973. Toxicity of acid coal mine spoils to plants. In ecology and Reclamation of Deviated Land. Vol. 1 Eds. R.J. Hutnick and G. Davis. Gordon and Breach, New York.

Bowen, H.J.M. 1978. Environmental chemistry of the elements. Acad. Press. New York. pp. 333.

Chadwick, M.J. 1973. Methods of assessment of acid colliery spoil as medium for plant growth. Ecology and Reclamation of Devasted Lands. Vol. 1 (Ed. by R.J. Hutnik and G. Davis), pp. 81- 91. Gardon & Breach, New York.

Chen, Y. and F.J. Stevenson 1986. Soil organic matter interactions with trace elements. In Y. Chen and Y. Avnimelech (ed.). The role of organic matter in modern agriculture p. 73- 119. Martinus Nijhoff Publ. Netherlands.

Coates, U.A. 1956. Colliery shale tips raised to grazing land. Surveyor, 115(3352), 21.7.1956, 527-9.

Cornwell, S.M. and Stone, E.L. 1973. Spoil type lithology and foliar composition of *Betula populifolia*. Ecology and Reclamation of Devastated Land. Vol. 1 (Ed. by R.J. Hutnik and G. Davis), pp. 105-20. Gordon and Breach, New York.

Cox, R.M. and T.C. Hutchinson 1980. The response of root acid phosphate acidity to heavy metal stress in tolerant and non- tolerant clones of two grass species. New Phytol. 87 : 359-364.

Down, C.G. 1975. Soil development on colliery waste tips in relation to age. J. Appl. Ecol. 12 : 613-635.

Elliot, C.G. 1955. The effect of temperature on chiasma frequency. Heredity, 9 : 385-398.

Fitter, A.H.; Handley, J.F.; Bradshaw, A.D. and Gemmell, R.P. 1974. Site variability in reclamation work. Landscape Design, 106 : 29-31.

Goodman, G.T. and Bray, S. 1975. Ecological aspects of the Reclamation of Derelict and Disturbed Land. N.E.R.C. Geo Abstracts Ltd. Norwich.

Gough, L.P. and R.C. Severson 1981. Biogeochemical variability of plants at native and altered sites, San Juan Basin, New Maxico. U.S. Geological Survey Professional paper 1134-D.

Imes, A.C. and M.K. Wali 1978. Government regulations of reclamation in the Western United States an ecological perspective. Reclamation Review, 1 : 75-78.

Indorante, S.J.; I.J. Jansen and C.W. Boast 1981. Surface mining and reclamation. Initial changes in soil character. J. Soil Water Conserv. 36(6) : 347-351.

Jackson, H.L. 1958. Soil chemical analysis. Prentice Hall Inc., Inc. Englewood, N.S., pp. 498.

Kabata-Pendias, A. and H. Pendias 1984. Trace elements in soil and plants. Boca Raton F.L. CRC Press.

Kimber, A.J; I.D. Pulford and H.J. Duncan 1978. Chemical variation and vegetation distribution on a coal waste tip. J. Apple. Ecol. 15 : 627-633.

Lindsay, W.L. 1974. Role of chelation in micronutrient availability. In : The plant root and its environment, ed. E.W. Carson, Charlottesville, Virginia : University Press.

O'Neill, R.V.; B.S. Ausmus; D.R. Jackson; R.I. VanHook; P. VanVoris; C. Washbowrne and A.P. Watson 1977. Monitoring terrestrial ecosystems by analysis on nutrient export. Air and Soil Pollut. 8 : 271-277.

Omodt, H.W.; F.W. Schroor and D.D. Patterson 1975. The properties of important agricultural soils as criteria for mined land reclamation. N.D. State Univ., Agric. Exp. Stn. Bull. 492.

Pahlsson, A.M. 1989. Toxicity of heavy metals (Zn, Cu, Cd, Pb) to vascular plants. A literature review. Water, Air and Soil Pollut. 47 : 287-319.

Parkinson, D. 1979. Microbes, mycorrhizal and mine spoil. pages 634-642 in M.K. Wali, editor. Ecology and coal resource development, Vol. 2, Perganen Press, New York.

Peach, M. 1965. Hydrogen ion activity. In : Methods in soil Analysis, Part-2, Agronomy 9, American Society of Agronomy, Madison, WI, pp. 922-923.

Ratha, D.S., G. Venkataraman and S. Pahala Kumar 1994. Soil contamination due to open cast mining in Goa : A Statistical approach. Environ. Technol. Vol. 15 : 853-862.

Richardson, J.R. and E.F. Greenwood 1967. Soil moisture in relation to plant colonisation in pit heaps. Proceedings of the University of Newcastle Philosophical Society, 1 : 129-136.

Schafer, W.M. 1984. Mine soil restoration and maturity. A guide for managing mine soil development. Symposium on surface coal mining and reclamation in the Great Plains, pp. 172-185.

Severson, R.C. and L.P. Gough 1983. Rehabilitation materials from surface coal mines in western USA : 1. Chemical characteristics of spoil and replaced cover soil. Reclaim. Revegetation Res. 2(2) : 83-102.

Sims, J.T. 1986. Soil pH effects on the distribution and plant availability of Manganese, copper and Zinc. Soil Sci. Am. J. Vol. 50, 367-373.

Singer, P.C. and W. Stumm 1970. Acidic mine drainage : the rate determining step. Sicence, (No. 4), 167 : 1121-1123.

Sparling, G.P., K.N. Whale and A.J. Ramsay 1985. Quantifying the contribution from the soil microbial biomass to the extractable p level of fresh and air dried soils. Australian J. of Soil Research 23 : 613-621.

Stach, E.; Mackowsky, M.T.; Teichmuller, M; Totlor, G.H.; Chandra, D. and Teichmuller, R. 1975. Stach's Textbook of coal petrology (2nd ed.), trans, by D.G. Murchison, G.H. Taylor and F. Zierke. Gebruder Borntrager, Berlin.

Tasker, A. and M.J. Chadvik 1978. The microdistribution of *Agrostis tenuis* on colliery spoil in relation to spoil chemical variability. J. of Appl. Ecol. 15 : 551-563.

Toy, T.J. and D. Shay 1987. Comparison of some soil properties on natural and reclaimed hill slopes. Soil Sci., 143 : 264-277.

USDA Hand Book 1977. Laboratory methods recommended for chemical analysis of mined-lands spoils and overburden in Western United States, US. Dept. Of Agriculture, Agriculture, Hand book No. 525, ed by Sandwal, P.M. and J.P. Power.

William, P.J. 1975. Investigation into nitrogn cycle in colliery spoil. The ecology of resources degradation and renewal (Ed. by M.J. Chadwick and G.T. Goodman), pp. 259-274. Blackwell Scientific Publications, Oxford.

CHAPTER - 25

INFLUENCE OF INDUSTRIAL SOIL POLLUTANTS ON VAM INFECTION, NODULATION AND GROWTH OF *ALBIZIA LEBBECK* (L.) BENTH

P. Srinivas, G. Laxma Reddy and S.M. Reddy

Microbiology Section, Department of Botany
Kakatiya University, Warangal - 506 009

INTRODUCTION

The success of the mutualism between the plant and fungus in a mycorrhizal relationship hinges in large part on carbon flow from plant to endophyte. Any disruption of that relationship has major consequences for both organisms, and any disrupting stress on a large spatial scale hypothetically has consequences for the community and ecosystem. This may be particularly true for the arbuscular mycorrhizal associations, which involve fungi that are ecologically obligate symbionts (Shafer and Schoeneberger, 1994). In recent years, the disposal of sewage water and industrial effluents on land has become a common practice in many countries. Sewage effluents are rich in organic wastes, while the industrial effluents contain toxic elements and heavy metals which cause undesirable changes in the soil ecosystem thus disturbing its physico-chemical setup and biological spectrum. The microbial communities, including VAM fungi may get affected by these effluents. Inspite of the significance of the problem, there is only limited information with regard to the effect of different effluents on VAM fungi (Reddy *et al.*, 1995; Raman *et al.*, 1995; Joseph *et al.*,

1995). In the present investigations, an attempt has been made to assess the influence of industrial effluents on VAM infection, nodulation and growth of *Albizzia lebbeck*, a multipurpose legume tree species at nursery level.

MATERIALS AND METHODS

The present investigations were carried out in the social forestry nursery situated in Kakatiya University Campus with sandy loam soil (pH 6.9 Ec 2.5 mhos/cm, organic matter 0.38 per cent, available nitrogen 250 kg/hectare, phosphorus 12.5 P_2O_5 Kg/hectare). Top soil of University botanical garden was collected in polythene bags (6 x 12 cm). Previously boiled and cooled seeds of *Albizzia lebbeck* were placed in polythene bags at the rate 5 per bag. The bags were regularly treated with different industrial effluents viz., A.P. Rayons, Nirup Synchrome Dye Factory, Azamzahi Textile Mills, Cement Pipe Company and Tannery factory. After the germination, the seedlings were thinned down to one for each bag. All the routine nursery precautions were taken. Simultaneous controls treated with water were maintained. After 90 days of growth, the saplings were taken out of the bag and various parameters like mycorrhizal infection, spore population, nodulation height of the plant, biomass, phosphorus and chlorophyll content were estimated. Percentage VAM infection was calculated by the formula given by Giovannetti & Mosse (1980) after clearing and staining the roots (Phillips & Hayman, 1970). Resting spore of VAM fungi were extracted by wet sieving and decanting method of Gerdeman and Nicolson (1963). Resting spores were identified by following the key provided by Schenck and Perez (1987). Phosphorus content in the oven dried shoot and root samples was estimated by molybdate-vandate method (Jackson, 1973). Arnon (1949) and Ikan (1969) formulae were employed to determine the content of chlorophyll-a, chlorophyll-b and carotenoids.

RESULTS AND DISCUSSION

From the results presented in Table 1, it is evident that the five industrial effluents under study exerted varied effect on the seedling growth of *A. lebbeck*. All the effluents under investigation has shown inhibitory effect on the mycorrhizal infection. However, the inhibitory effect varied with the type of effluent. Dye factory and tanney effluents followed by textile mill and cement pipes company effluents have shown more adverse effect on the mycorrhizal infection. The inhibitory effect of dye factory and tannery effluents may be due to the presence tannins and other heavy metals like chromium in the effluents. Rayons factory effluents followed by cement factory effluents have shown less inhibitory effect when compared to other various effluents. Dye factory and tannery effluents had exerted more inhibitory effect on the height and biomass production of the sedlings. This inhibitory effect may be direct due to effect of effluents on plant or indirect through the inhibition of mycorrhizal association and nodule formation. Chlorophyll (a,b) contents were drastically affected by cement pipe company effluents. Interestingly, dye factory and tannery effluents which had shown adverse effect on symbiotic associations and biomass, exhibited very less inhibitory effect on the chlorophyll synthesis. Phosphorus absorption was drastically affected by cement pipe company effluents. In general Rayon factory effluents exerted least inhibitory effect on the chlorophyll synthesis and phosphorus uptake.

Table 1. VAM Colonization, growth and biomass production of *A. lebbeck* under the influence of five industrial effluents.

					A		B		C	
1	2	3	4	5	6	7	8	9	10	11
A.P. Rayons (Kamalapur)	26	112	32	27.5	19.0	12.5	0.78	0.64	0.78	0.61
Nirup Synchrome dye factory (Waddepally)	18	88	16	18.6	10.0	6.5	0.74	0.56	0.84	0.46
Tannery Industry (Desaipet)	18	76	18	19.5	16.5	12.8	0.88	0.59	0.48	0.37
Ajamjahi Textile Mill (Warangal)	24	94	30	23.0	22.0	14.0	0.68	0.53	0.68	0.54
Cement Pipes Company (Gopalapuram)	24	98	24	25.0	24.0	16.5	0.51	0.43	0.59	0.33
Control	48	138	46	38.5	28.0	19.0	0.91	0.66	0.98	0.78

1= Name of the effluent, 2= Percentage of infection, 3 = No. of Propagules/100 grams, 4=Nodulation, 5 = Height of the plant (cm), A = Biomass, 6 = Fresh wt.(gm), 7=Dry wt. (gm), B= Chlorophyll, 8=a, 9=b, C= Phosphorus content, 10=Root mg/plant, 11= Shoot mg/ plant.

Critical perusal of Table 2 reveals that the five common VA mycorrhizal genera viz., *Glomus, Sclerocystis, Gigaspora, Acaulospora* and *Scutellospora* were associated with the seedlings treated with all the five industrial effluents. However, the percentage of incidence varied with the type of effluent. In all the treatments, resting spores of *Glomus* dominated followed by *Gigaspora* and *Scutellospora*. Different genera responded differently towards effluents. Though *Glomus* species dominated, its spore number decreased in all the treatments over the control. *Sclerocystis* spore population was enhanced in dye factory, tannery and texile mill effluents, whereas Rayons factory and cement factory effluents inhibited spore proliferation. All the five of industrial effluents have shown stimulatory effect on the spore population of *Gigaspora* and *Acaulospora*. Rayon factory, textile mill and cement pipe company effluents have shown stimulatory effect whereas dye factory and tannery effluents exerted inhibitory effect on *Scutellospora*. A number of factors are attributed to the selective effect of various effluents on the spore population of VAM fungi. Abbott and Robson (1977) reported that neutral to slightly alkaline soil is known to favour the predominance of Glomous species, but reduction in soil pH due to effluent irrigation did not markedly alter the effluents. Apart from toxic heavy metals, a number of factors such as pH, BOD, COD organic matter content and other essential and non-essential elements may be responsible for the observed inhibitory effect. In contrast to present studies Raman *et al*. (1995) reported positive effect of tannery effluents on 19 plants growing in polluted soil. Similarly Babu *et al.*, (1995) and Joseph *et al.*, (1995) observed no inhibitory of distillary sludge and petro-effluents on the plant growth. Kehri and Chandra (1990) have also reported sewage irrigation was favourable for mycorrhiza formation in crops studied by them. On the otherhand, toxic effect of sewage and sludge on mycorrhizal fungi has been reported by Lambert (1985).

Table 2. Incidence of VAM fungal genera under the influence of different industrial effluents.

1	2	3	4	5	6	7
A.P. Rayons	112	75.0	4.4	8.0	7.1	5.3
Nirup Synchrome dye factory	118	78.0	5.9	7.6	3.3	4.2
Tannery Industry	126	77.0	6.3	8.7	3.9	3.1
Ajamjahi Mill Textile Company	94	63.0	10.6	7.4	8.5	9.5
Cement Pipe Company	98	73.0	4.0	10.2	5.1	7.1
Control	136	84.5	5.1	2.9	4.4	2.9

1= Name of the effluent 2=No. of propogules/100 gram soil, 3= *Glomus* species, 4= *Sclerocystis* species, 5= *Gigaspora* species, 6= *Scutellospora* species, 7= *Acaulospora* species

Spore density in the rhizosphere soil of seedling treated with different effluents varied with the type of effluents. In general all the five effluents under investigation had shown inhibitory effect on the mycorrhizal resting spore population. Similar to mycorrhizal colonization, resting spore population was more adversely affected by dye factory and tannery effluents. The observed variation in the spore densities in different host plants may be due to the indirect effect of effluents on the host plant growth that in turn effected the fungal symbiont. A direct correlation could be observed between the mycorrhizal colonization and resting spore population. Srinivasan *et al.,* (1996) reported that Rayon factory effluents significantly reduced spore densities. Changes in edaphic factors like pH, N, P, Ca, Zn and Mn may also influence VAM fungal sporulation (Hayman, 1982, Thompson, 1986 and Boyle and Paul, 1988). The present industrial effluents had also shown inhibitory effect on the nodulation of *A. lebbeck*. Though the suppressing effect of various effluents varied, the inhibitory effect in mycorrhizal infection was almost same.

Various growth parameters such as height, biomass, chlorophyll and phosphorus content were also adversely affected with the treatment of diversity of indigenous VAM flora. VAM fungal species are reported to develop strains adopted to varying soil conditions. *Gigaspora candida* and *G. multicaule* were exclusively found in Rayon effluent polluted soils (Srinivasan *et al.,* 1996). They further reported that the number of VAM fungal species were more in effluent irrigated soils than in ground water irrigated soils.

REFERENCES

Abbott, L.K. and A.D. Robson 1977. Growth stimulation of subterranean clover with VA mycorrhizas. *Aust. J. Agric. Res.,* 28 : 639-649.

Arnon, D.J. 1949. Copper enzymes in isolated chloroplast polyphenol oxidase in *beta vulgaris*. *Plant Physiol.,* 24 : 1-15.

Babu, S.R., V.C. Saralabai and M. Vivekanandan, 1995. Vesicular arbuscular mycorrhizal colonization and biomass production in economically important crop plants grown directly over the sludge. proc. Third Nat. Con. on Mycorrhiza, pp. 159-163.

Boyle, M. and E.A. Paul, 1988. *Soil Biol. Biochem.* 20 : 445.

Gerdemann, J.W. and Nicolson, 1963. Spore of mycorrhizal endogone species extracted from soil by wet sieving and decanting. *Trans. Br. Mycol. Soc.* 46 : 489-500.

Hayman D.S. 1982. Influence of soils and fertility and survival of vesicular arbuscular mycorrhizae. *Phytopathology* 72 : 1119- 1125.

Ikan, R. 1969. Natural products. A laboratory guide, Academic Press, New York, pp. 101.

Jackson, M.L. 1973. Soil Chemical Analysis. New Delhi : Prentice-Hall, India.

Joseph Udhaya, I.L. Kothari and Nimisha Patel 1995. Effect of mycorrhization on two crops raised on land disturbed by petro-effluents. Proc. 3rd Nat.Con. on mycorrhiza. pp.164-167.

Kehri, H.K. and S. Chandra, 1990. Mycorrhizal association in crops under sewage farming. *J. Indian Bot. Soc.* 69:267-269.

Lambert D.H. 1985. Proc. 5th N. Ame. Con Mycorrhizae USA.

Phillips, J.M. and D.S. Haymann, 1970. Improved procedure for clearing roots and staining parasitic and VAM fungi for rapid assessment of infection. *Trans. Br. Mycol. Soc.* 55 : 158-161.

Rama Pulla Reddy P., P. Jagan Mohan Reddy, P. Prakash, and C. Manoharachary, 1995. Association of vesicular-arbuscular mycorrhizal fungi in soil polluted with industrial and sewage effluents *Proc. 3rd Nat. Con. on Mycorrhiza*, pp. 155-158.

Raman, N., K. Sambandan, C. Sahadevan, and T. Selvaraj, 1995. Distribution of vesicular arbuscular mycorrhizal fungi in tannery effluent polluted soils of Tamilnadu, India. Proc. of the 3rd Nat. Con. on Mycorrhiza. pp. 168-178.

Schenck, N.C. and Y. Perez 1987. Manual for the identification of VA mycorrhizal fungi. In VAM Florida University, Gainesville, U.S.A., 245 pp.

Shafer, S.R. and M.M. Schoeneberger 1994. Air pollution and ecosystem health. The mycorrhizal connection. In Mycorrhizae and plant health Ed. F.L. Pfleger and R.G. Linderman. APS Press.

Srinivasan, D., T. Muthukumar and K. Udhaya 1996. Influence of effluent irrigation on VA mycorrhizae. *Indian J. Microbiol.* 36 : 37-40.

Thompson, J.B. 1986. Soil less cultures of vesicular mycorrhizae of cereals. Effect of nutrient concentration and nitrogen source. *Can. J. Bot.* 64 : 2282-2294.

CHAPTER - 26

ROLE OF CYANOBACTERIA IN RECLAMATION OF 'USAR' LAND

R.K. Gupta, K.D. Pandey* & A.K. Kashyap*

Department of Botany, Koshi College, Khagaria

*Department of Botany, Banaras Hindu University, Varanasi - 221 005

INTRODUCTION

Depending on the salt present, the 'Usar' soils have been classified into saline, saline alkali and alkali soils. The alkali soils, which have been extensively studied in various parts of India, are characterized by exchangeable sodium more then 15% (Dhar and Mukharji, 1936; Singh, 1961), low quantities of sodium salts and pH values usually ranging between 8.5 to 10.5.

The physical and chemical methods such as excessive irrigation, crop rotation, application of gypsum and calcium chloride, presently used in reclamation of 'Usar' soils, do not help complete removal of soluble salts and exchangeable sodium. Describing a biological method, Singh (1950) for the first time suggested the significant role of cyanobacteria in the reclamation of 'Usar' soil in India however, natural recuperation of soil fertility year after year is predominantly an algal process. It is believed that the extracellular products secreted in soil by these algae lower the pH of surrounding medium and consequently help in increasing the availability of minerals to plants. In view of the above the present study was planned to investigate the changes in 'Usar' soil following the growth of *Cyanobacteria Nostoc calciocola* in natural as well as in laboratory.

MATERIALS AND METHODS

Collection of soil samples

Soil samples were collected from 'Usar' land at depths of 0-10 cm and 10-20 cm in month of June when no growth of cyanobacteria was noticed on the soil surface. The soil samples were dried and powdered. After the first shower of rains, collection of soil samples were made at 15 days interval over a period of 4 months The algal forms in soil were separated mechanically and examined under microscope. The soil samples with algal crusts were than air dried and powdered.

Culture conditions for the growth of cyanobacteria

Nostoc calcicola, a N_2-fixing blue-green alga, isolated from 'Usar' soil sample was cultivated in Allen-Arnon medium supplemented with A_5 trace element (Allen & Arnon, 1955). The cultures were grown photoautotrophically in a culture room at 251°C and illuminated with cool florescent light (approximately 1700 lx) on the bottom of the culture vessels.

Effect of cyanobacterial growth on 'Usar' soil

Three types of experiments were performed to evaluate the role of cyanobacteria in influencing physico-chemical properties of the 'Usar' soils. The first experiment was confined to marked protected areas in field conditions. The second one, 'Usar' soil patches with thick growth of cyanobacteria were brought to the laboratory, flooded with double distilled water and incubated for two years in natural conditions. The third set of experiments were performed in the laboratory under standard growth conditions.

To a 100 gm of 'Usar' soil sample in 250 ml culture vessel, 150 ml distilled water was added and sterilized at 15 lb. pressure/inch for 15 min. Exponentially growing blue-green alga *N. calcicola* was incubated in these flasks except in the control. All these flasks were incubated under photoautotrophic growth conditions in a culture room.

Quantitative estimation of cyanobacteria

Soil blocks of one cm^2 were taken randomly from the petriplates containing 'Usar' soil and algae. The algal mass was scrapped, separated to the possible extent and was suspended in the culture medium. The flasks containing algal samples were shaken on a gravatory shaker to separate the algal growth into individual filaments. Quantitative estimations of the algae were made by plating technique.

Effect of *N. calcicola* and pyrite on 'Usar' soil

A 40 gm powdered soil sample was taken in petriplate, spread into one cm thick soil horizon and it was thoroughly mixed with pyrite (FeS_2), the quantities of which ranged from 25 to 250 mg/100 gm 'Usar' soil. Two sets of experiments were performed. In one set, the petriplates with soil and pyrite were sterilized but the other set was not sterilized. After flooding with double distilled water, the plates were inoculated with exponentially growing *N. calcicola* and incubated in a culture room under standard growth conditions for 12 months. The soil of each petriplate was flooded with sterilized distilled water at 7 days intervals.

Effect of 'Usar' soil extract on growth of *N. calcicola*

Samples of 'Usar' soil and double distilled water (1:10 w/v) taken in one litre capacity flasks were steam sterilized thrice at 24 h interval and finally filtered and stored. Soil extract of a given percentage along with Allen-Arnon medium was supplied to observe its effect on the algal growth. Pyrite (199 g/ml) was added to the above combinations to see their combined effect on algal growth. The level of growth was measured in terms of protein, by using the Lowry method as modified by Herbert *et al.*, 1971.

Soil analysis

Texture, water holding capacity and pH of the soil sample were determined by following the methods as described by Piper, 1966. The soil pH was measured in a soil water suspension (1:5 w/v) with a pH meter. Organic carbon content (%) and total nitrogen in soil samples were determined by following Walkley and Blacks rapid titration method and microkjaldahl techniques, respectively (Piper, 1966). Exchangeable Ca^{++} and Na^{+} (m.e. %) in soil samples were determined after leaching the soil with ammonium acetate and titrating the lechate against sulphuric acid (0.1 N) and potassium permagnate after the addition of sulphuric acid, respectively as in the method described by Jackson, 1962. Total phosphorus was determined colorimetrically after leaching it with ammonium acetate. The organic phosphorus was estimated by using the ignition method as described by Jackson, 1962. Available phosphorus was determined colorimetrically by the method of Olsen *et al.*, 1954.

RESULTS

There were horizons A1 and A2 that were investigated in 'Usar' land soil profile. The horizon A1 (0-10 cm) comprised small to large clods. The soil particles were loose and the colour varied from whitish to grey. The soil texture was clay type and pH was 10.5. Horizon A2 (10-20 cm) was more calcarious with compact soil and particle colour varied from whitish grey to deep grey. The values of clay percentage, pH and water holding capacity were higher in A2 than A1 horizon (Table 1).

Table 1. Physical and chemical properties of "Usar" soil.

Properties	(Soil Profile (depth cm)	
	A_1 (0-8)	A_2 (10=20)
Colour Compactness	White Soil Particles loosely Arranged	Slightly grey Particles more compact
pH	10.2	10.5
Clay*	18.85	25.12
Water holding capacity*	41.30	47.01
Na_2CO_3*	0.025	0.045
Organic carbon*	0.15	0.10
Total Nitrogen*	0.028	0.010
C/N ratio	5.35	10.0
Exchangeable (m.e.%)		
Ca	3.95	3.03
Na	6.92	8.55
Ca/Na ratio	0.55	0.35
Phosphorus (mg/100g)		
Total P	30.05	31.85

On dry weight basis

The chemical analyses of control soil samples of horizons A1 and A2 with no apparent growth of blue-green algae showed variations on the dry weight basis in percentage sodium carbonate from 0.025 to 0.045 of organic carbon from 0.10 to 0.15 and of nitrogen from 0.028 to 0.01. The ratio of exchangeable sodium and calcium in horizon A1 were higher than horizon A2. The contents of phosphorus was almost similar in the two horizons.

The algal flora observed during rainy season included several non-N_2-fixing species (*Oscillatoria* sps., *Plectonema* sp. and several species of *Phormidium*) and N_2-fixing species (*Scytonema javanicum, S. hofmanii, Anabaena ambigua, A. variabilis, Cylindrospermum* sps. *Nostoc muscorum, N. commune, N. linckia* and *N. calcicola*) of blue-green algae. *Aulosira fertilissima* forms a thick mat and was a dominant species in algal collection throughout the rainy season.

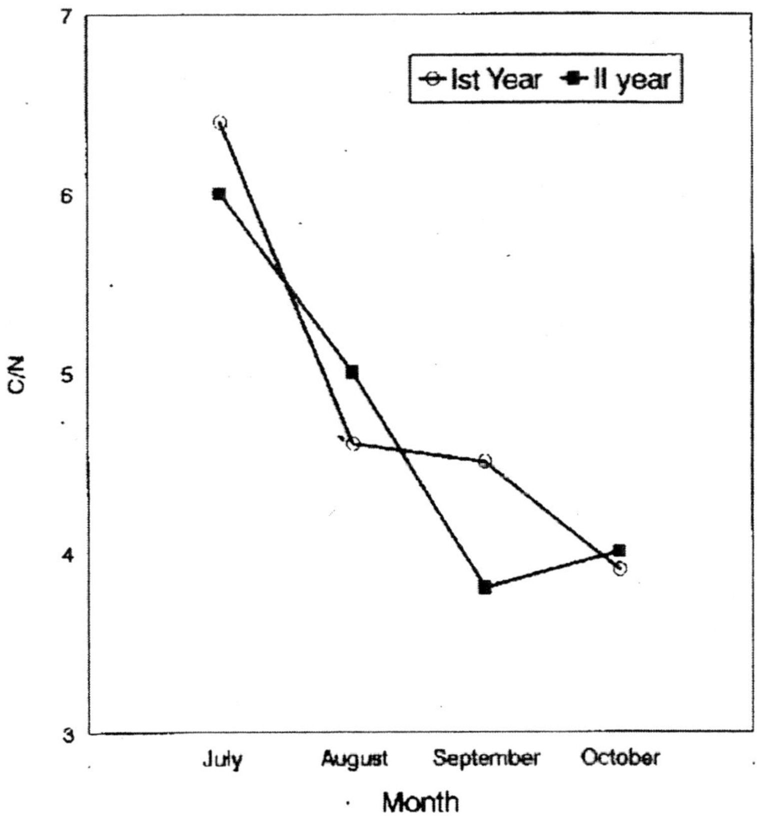

Fig. 1. Effect of cyanobacterial growth on C/N ratio in two successive years.

The data collected in rainy season of two consecutive years are summarized in Fig. 1. The C/N ratio in rainy season decreased due to growth of blue-green algae.

The experiments conducted with 'Usar' soils in glass through under natural conditions showed decrease in pH from 10.5 to 8.2 and increase in water holding capacity, total N-content and organic carbon. C/N ratio decreases, also decrease in sodium was observed. With respect to phosphorus there was increase in the total phosphorus (Table 2).

Table 2. Changes in "Usar" soil characters due to Cyanobacterial growth after the rainy season

Soil properties	Pre phase growth	Post phase growth
pH	10.5	8.2
Total Nitrogen**	0.027	0.165
Oraganic Carbon**	0.162	0.658
C/N ratio	6.0	3.07
Na	8.05	6.27
Total P (mg/100 gm soil)	30.15	32.35

* Algal Flora includes both N_2 and non N_2=fixing Cyanobacteria
** % On dry weight basis

As a result of the growth of *N. calcicola* in almost pure culture on 'Usar' soil for a period of two years, the soil pH decreased from an initial value of 10.5 to 7.8 (Table 3), total nitrogen increased nearly five fold and organic carbon increased from 0.14 to 0.58%. The Ca/Na ratio showed a decrease while the percentage of total phosphorus increased.

Table 3. Changes in "Usar" soil after the growth of *Nostoc calcicola*

Properties	Pre phase growth	Post phase growth
pH	10.5	7.8
Total Nitrogen**	0.025	0.118
Oraganic Carbon**	0.14	0.582
C/N ratio	5.6	4.09
Na	8.02	6.05
Total P (mg/100 gm soil)	30.15	32.19

** % On dry weight basis

The effect of pyrite (FeS_2) on the algal population and on the growth of blue-green alga *Nostoc calcicola* and effect of pyrite and *N. calcicola* on the pH of 'Uaer' soil were studied under sterilized and non-sterilized soil conditions. The pyrite at all concentrations was found to increase the algal population. The population increased with gradual increase of pyrite upto the level of 200 mg/100 gm sterilized soil and 175 mg/100 gm non- sterilized soil (Fig. 2). Further increase in pyrite concentration tended to decrease the algal population. The pyrite induced algal population together with pyrite helped to bring the soil pH towards the neutral (Fig. 3). The growth of *N. calcicola* was increased with 'Usar' soil extract with or without the addition of pyrite (100 g/ml) into it. The growth measured in terms of g protein/ml culture, gradually increased with increase of 'Usar' soil extract (%) in the medium upto the extent of 40% (Fig. 4) the growth being maximum in both pyrite and non-pyrite supplemented conditions. An increase of soil extract from 40-60% in the growth medium resulted in the decrease of algal growth which was, however, still better than that of control. Further increase of soil extract in the medium led to inhibition of growth of the alga (Fig. 4).

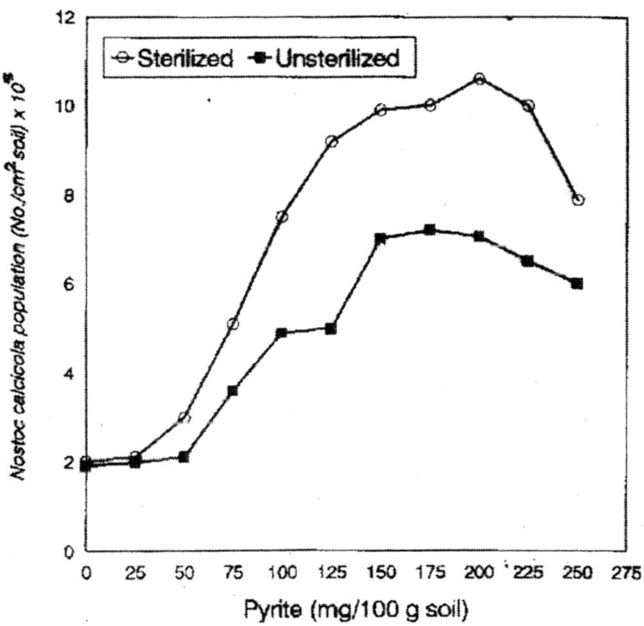

Fig. 2. Effect of Pyrite on *Nostoc calcicola* population of the sterilized (open symbol) & unsterilized (closed symbol) 'Usar' soil.

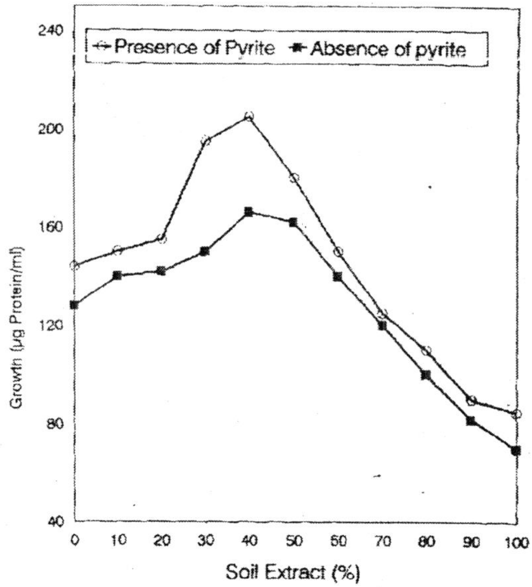

Fig. 3. Combined effect of Pyrite & *Nostoc calcicola* population on pH of the sterilized (open symbol) & unsterilized (closed symbols).

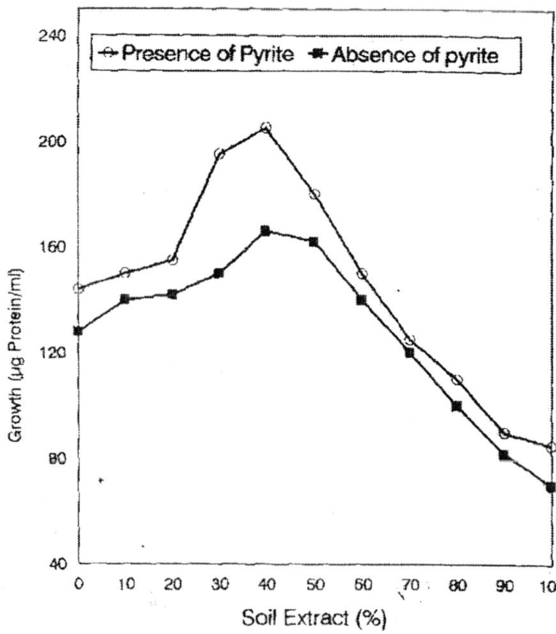

Fig. 4. Effect of 'Usar' soil extract on the growth of Nostoc calcicola in presence and absence of Pyrite.

DISCUSSION

It is suggested that 'Usar' development is due to gradual sodiumization resulting in the degradation and pH increase of soil (Singh, 1961). The pH is an important factor which determines the algal flora in terms of its composition occurrence and dominance on a given soil surface. It is seen that algal forms of acidic and alkaline soil differ. For example, the growth of green algae is favoured by low pH and that of blue-green algae by higher pH values (Pandey, 1965). Figg, (1956) reported that in natural conditions most of the cyanobacteria grow in neutral to alkaline conditions and that the growth of N_2-fixing blue-green algae in rice field is most commonly limited by low pH (Fogg *et al.,* 1973). The 'Usar' soil having higher pH value and sodium content favours the growth of N_2-fixing blue-green algae and the decrease in pH consequent to the growth of algae on such soil may be considered as a reclamation process of 'Usar' soil by blue- green algae. Sodium is required as an essential macronutient for the growth and enhanced the N_2-fixation in blue-green alga (Apte and Thomas, 1980; Batherton and Van Baalen, 1971). The experiments conducted under field and labortory conditions reveal that *N. calcicola* could tolerate high salt concentrations and alkaline pH and during its growth, the pH value of the 'Usar' soil was gradually reduced. Pyrite alone which is used in the chemical method for reclamation of 'Usar' land, may cause toxicity by releasing ferrous and sulfide ions in the soil. However, blue-green algae alone grow on soil with pyrite can utilize for their better growth and reclaming the 'Usar' soil on the other hand. It is reported that as a result of successive algal inoculation there is a significant reductin in the contents of oxidizable matter total sulfides and ferrous iron in the soil (Aiyer *et al.*, 1972).

In a long term experiment, an increase in organic carbon was noted following the inoculation of blue-green algae. However, it became steady at the end of three year period (Sankaran, 1971). Singh, 1961 reported an increase in organic matter after the growth of cyanobacteria in 'Usar' soil. The bulk of organic matter produced by algal growth remained in the soil to decompose and build-up the organic N content which subsequently maintained the fertility of soil through the year (Roger and Kulasooriya, 1980). The amount of N_2-fixed is estimated to be 80 kg/ha in paddy field by cyanobacteria. According to Watanebe *et al.*, 1979 per crop. The N_2-fixation by blue-green algae seems to be comparatively rapid process than the organic carbon assimilation in 'Usar' soil, hence a decrease in C/N ratio has been observed.

REFERENCES

Aiyer R.S. Salajudeen S., Venkataraman G.S. 1972. Long term algalization field trial with highyielding variteis of rice (*Oryza sativa* L). Indian J. Agril. Sci 42:380-383.

Allen M.B. and Arnon, D.I. 1965. Studies on nitrogen fixing blue-green algae. I. Growth and nitrogen fixation by *Anabaena cylindrica* Lemm. Plant Physiol. 30:366-372.

Apte S.K. Thomas J. 1980. Sodium is required for nitrogenase activity in cyanobacteria. Current Microbiol. 3:291-293.

Batterton J.C, Van ballen C. 1971. Growth responses of blue- green algae to sodium chloride concentration. Arch. Microbiol. 76:151-165.

Dhar N.R., Mukherjee S.K. 1936. Alkali soils and their recamation. Proc. Nat. Acad. Sci. India. 6:136-148.

Fogg G.E. 1956. Nitrogen fixation by photosynthetic organism. Ann. Rev. Plant Physiol. 7:51-70.

Fogg. G.E. Stewart W.D.P., Pay P. and Walsby A.E. 1973. The blue-green algae. Academic Press, London.

Jackson M.L. 1962. Soil chemical analyses. Prantice Hall. N.J.

Lowry, O.H., Rosenbrough, N.J., Faor A.L. and Ramdall, R.J. 1951. Protein measurement with phenol phenol reagent. J. Biol. Chem. 193:265-275.

Olsen S.R. Cole C.V., Frant, S. and Dean L.A. 1954. Estimation of available phosphorus in soils by extraction with sodium bicarbonate, cere 939 (Us Dept. Agril.)

Pandey D.C. 1965. A study of the algae from paddy soils of Ballia and Ghazipur districts of Uttar Pradesh. India. II (A) Taxonomic considerations-Cyanophyceae. Nova Hadwigia Z. Kryptogamenk 10:177-209.

Piper C.S. 1966. Soil and plant analyses. Inter Science Publ. Inc. New York.

Roger P.A. and Kulasooriya S.A. 1980. Blue-green algae and Rice. The International Rice Research Institute. Los banos, Philippines.

Sankaram, A. 1971. Work done on blue-green algaein relation to agriculture. Indian Council of Agril. Res. New Delhi.

Singh R.N. 1950. Reclamation of 'Usar' lands in India through blue-green algae. Nature. 165:325-326.

Singh R.N. 1961. Role of blue-green algae in Nitrogen economy of Indian Agriculture. Indian Council of Agril. Res. New Delhi.

Watanable I., Lee K.K. Alimagno B.V., Sato M., Del Rosario D.C. and De Guzman M.R. 1977. Biological N_2 fixation in Paddy field studied by in-situ acetylene reduction assay. International Rice Res. Inst. Res. Paper Ser. 3:1-16.

CHAPTER - 27

EFFECT OF LIMING ON SOME HEAVY METALS OF COPPER MINE TAILING AMENDED SOIL OF KHETRI AREA

[1]Pankaj Kumar and [2]Niraj Kumar

[1]NTPC, Kahalgoan, [2] University Department of Botany
T.M. Bhagalpur University.

INTRODUCTION

Hindustan Copper Limited is engaged in producing finished Copper metal with an annual capacity of approx. 3100 million Ton at Khetri Copper Complex Rojasthan. About 85% ore after its processing is rejected as a waste meterial called TAILINGS. Tailings obtained from copper mining at Khetri, are highly acidic, deficient in terms of availability indexes of N, P and K and they possess high amount of heavy metals. These tailings when brought to the field cause heavy damage to the soil environment creating unfavourable condition for plant growth. Soil adjoining mining area are near neutral (pH = 7.2). Mixing up of tailings with soil gives fairly acidic mixture. To remove the acidity of the soil or soil mixtures, commonly used materials are oxides, hydroxides and carbonates of Ca and Mg (Pearson and Adams, 1967). They are relatively inexpensive and leave no residue in the soil (Barber, 1967). In the present study CaO (reagent Grade) has been chosen for neutralising the acidity of S-T Mixtures. Present study attempts to assess the effects of liming on different S-T mixtures.

Mine tailings contain excessive concentration of various heavy metals that are toxic to plant, (Bradshaw, 1952). Lime addition to soil drastically modify the solubilities of numbers of the trace elements (Coleman *et al.*, 1958). Liming of a soil often affects the avilability of heavy metals to the plant (Smith, 1968 and Singh, 1971). It is reasonably well understood that the availability of all micronutrients except Mo decreses with pH. It was reported that the solubilites and availability of Cu, Zn and Fe are generally decreases when soils are limed (Gupta, 1970). It was also found that soil pH significant affected solubility of Fe, Cu and Zu cations and can be expressed as a function of soil pH (Sinha, 1978). The decreasing trend of Zn in soil after liming was reported by numbers of authors (Bandopadhaya & Adhikari, 1975, Chatterjee & Dass, 1964, Ganjir *et at.*, 1973 & Iyengar *et al.*, 1981). Rai and Mishra (1967); Grewal *et al.*, (1969); observed a decreased in the amount of exchangeable Copper with and increase in $CaCO_3$ content. MaClaren *et al.*, (1973) found increased Cu adsorption over a pH range of 5 to 7. Availability of Cu is dependent on soil pH but does not normally increases appreciably unless pH falls below 5.0. Liming significantly reduced the available Fe content in soils (Gupta *et al.*, 1970)

MATERIALS AND METHODS

Soil samples used in this experimental work were collected from fallow land 3-4 km away from Khetri Copper Complex. Soil samples were taken from 0 to 9 inches of the top soil. Tailings samples were collected form the Tailings ponds at Khetri Copper Complex from upper 0-12 inches. Then the soil Tailings were air dried, powerded and sieved through a 2mm sieve and were analysed (Table 1). Soil and Tailings were mixed in different proportions by weight and their lime requirement were determined. Soil and Tailings were mixed in 25%, 33%, 50%, 66%, 75% and 100% Tailings ratio. Lime requirement of the soil-tailing mixture were determined by the procedure given by Shoemaker *et al.,* (1961). Once the lime requirement in the form of gm of CaO/500 gm of sample of bring the sample to pH 7 was known, lime was added of different S-T Mixtures. Three different doses of lime added to one particular S-T Mixture.

(A) Half of the lime requirement

(b) Actual lime requirement

(c) Double the lime requirement

It can be represented as A1, B1, C1 (100% T in each) or A2,B2,C2 (75% T in each) .... A6,B6,C6 (25% T in each)

Table 1. Analysis of Soil Tailings

pH	7.12 -3.95	EC	127.4-2.88 mhos/cm
Organic Carbon	0.176-0.083%	N (Available)	72-23.6 ppm
N (Total)	0.098-0.021%	PO_4 (Available)	16.36-1.2 ppm
PO_4 (Total)	296-225 ppm	K (Available)	110-25 ppm
K (Total)	1780 -1700 ppm	Fe (Total)	0.02956- 0.1494 %
Cu (Total)	160-1560 ppm	Zn (Total)	57.6-120 ppm
Fe (Available)	25.6-212.8 ppm	Cu (Available)	8.8-183.8 ppm
Zn (Available)	2-14 ppm		

All the experiments were conduced at 50% water holding capacity of the corresponding S-T Mixtures and at room temperature, pH of S - T Mixtures were determined by digital pH Meter and Available heavy metal analysis was carried out by AAS by DTPA Extraction methods.

RESULT AND DISCUSSION

Periodic changes in pH of different S-T mixtures with three doses of lime addition at the intervals of 15, 30, 45 and 60 days are shown in Table 2. Application of lime resulted in raising of pH for all doses of lime addition and for all S - T Mixtures. The pH values were found to be less than expected after different amount of lime additions except in the case of 50°C and 66B, where the increase in pH due to liming was appropriate. From 15 days to 60 days pH of all S - T mixtures for all doses of lime additions showed decreasing pattern after 30 days instead of 15 days. Decrease in pH in S - T mixtures high in soil % was similar and uniform.

All observed periodic variations may be explained by taking into consideration of the fact that tailings obtained from Copper mining contained sulphides of various heavy metals which released sulphuric acid due to oxidation of these sulphides to sulphates (Pulford *et al.*, 1983 and Ivarson, 1973). Thus oxidative dissolution of sulphides minerals in commonly associated with an increasing acidification of surrounding medium and so predicated lime requirement were found to be less than the actual.

Table 2. Periodic changes in pH of S - T Mixtures due to liming

S - T Mixtures	Time (Days)				
	0	15	30	45	60
25A		5.6	5.3	5.1	5.0
25B	4.96	6.5	6.3	6.2	6.1
25C		7.2	7.1	7.0	7.0
33A		5.4	5.2	5.1	5.0
33B	4.79	6.6	6.2	6.1	6.1
33C		7.4	7.2	7.2	7.1
50A		5.6	5.4	5.3	5.2
50B	4.39	6.9	6.5	6.3	6.2
50C		7.7	7.2	7.0	6.9
66A		5.5	5.4	5.3	5.2
66B	4.30	7.1	7.3	7.2	7.0
66C		7.5	7.4	7.2	7.1
75A		5.5	5.6	5.3	5.2
75B	1.14	6.8	7.1	7.0	7.2
75C		7.6	7.4	7.4	7.3
100A		5.9	5.6	5.5	5.4
100B	3.95	6.7	7.0	6.8	6.8
100C		7.3	7.1	6.9	6.8

Other observation of changes in pH, irregularities in trends marked differences in the uniformity of the trends between mixtures high in soil % and those or tailing percentage, may be attributed to the fact that tailing do not have the buffering Capacity as soils and tailing represent extremely, complex and environment. So a relatively small change in lime addition on S - T Mixtures high in tailing % can change a number of variables which controls pH and hence fluctuations and irregularities can not be overruled.

Availability of Cu in S - T Mixtures decreased following liming for all doses of lime additions (Table 3) and decrease was more in case in lime addition B and C. Thus, all periodic variations of Cu availability may be correlated with pH. As the pH increase Cu^{3+} from solution is likely to precipitate as $Cu(OH)_2$ and decrease in Cu availability with increasing pH is probably due to the preference of soil exchange sites of hydrolysis product of this actual. The periodic variations and the decrease in Cu availability following liming may be accounted for the hydrolysis of Cu^{3+} and removal of Al^{3+} and H^{+} from exchanges sites & adsorption of Cu on Fe and Al hydroxides (Kinniberg *et al.*, 1976). The increase in Cu availability in case of 75% mixture and tailing above pH 7 may be attributed to soluble organic complexing agents (Saeed and Fox, 1977).

Table 3. Periodic changes in Available Cu (ppm) of S - T Mixtures due to liming.

S - T Mixtures	Time (Days)				
	0	15	30	45	60
25A		40.4	46.2	48.4	51
25B	56.8	25.6	22	23	23
25C		20.6	19.2	21	24
33A		52.2	54.6	55	55
33B	79.8	28	28.4	30.2	28.2
33C		20.2	22	25.6	20
50A		68	72.4	66.2	60.8
50B	109.6	35	39	38	36
50C		25.4	30.6	27.8	24.4
66A		70	65	65	70.4
66B	132.9	32.4	33	33	30.4
66C		26.4	30.4	31	30
75A		50.2	43	40.6	40.8
75B	137	14.2	20.4	15	15
75C		21.02	21	19	19
100A		65.8	57	50.2	50.8
100B	183	23	20.6	20.6	18.8
100C		28.4	23	20.8	23

Availability of Zn in different S - T Mixtures and their periodic variations have been shown in Table 4. Liming of all S - T Mixtures reduced the concentration of soluble Zn for all S - T Mixtures. Reduction in available Zn was found very much effective in high tailing and mixtures. Decrease in availability of Zn with pH can be observed from Fig. 2. We find that on liming the Zn availability decreases with pH w.r.t. to time. Thus the effect of liming on Zn availability is maintained over time.

Table 4. Periodic changes in Available Zn (ppn) of S - T Mixtures due to liming.

S - T Mixtures	Time (Days)				
	0	15	30	45	60
25A		5	5	5.5	5.6
25B	6	3	2.8	3.2	3.2
25C		3	2.5	3	2.8
33A		4	4.8	4.5	5.2
33B	4	3	4	4	4.5
33C		2	2	3	3.2
50A		6.2	5.5	5.5	5.8
50B	10	4	4.7	4.9	5
50C		2	3.2	3	3
66A		6	5.8	5.5	5.9
66B	10	3	2.4	2.7	2.5
66C		2.5	2.8	3.1	3.3
75A		4	3.5	4	3.8
75B	10	2	2	2.5	3
75C		2	2	2	2
100A		4	3.6	4	4.2
100B	14	2	2.4	3	2.5
100C		2	2	2.5	2.5

Decrease in Zn solubility with increasing pH may be accounted for by generation of pH dependent charge sites on organics or clays (Bache, 1976). Zn absorption on Fe and Al hydroxide surfaces may be responsible for decrease in Zn solubility with increasing pH in soil and soil like system (Kinniberg *et al.*, 1976) and we know that adsorbed Zn^{3+} is the exchangeable form in soil. Another factor which also might be playing role in the solubility of Zn in soil may be the hydrolysis of Zn^{3+} ion into its hydroxide with the development of alkalinity in soil and soil like system after liming.

Fe availability decreased by liming and decrease in available Fe was more with higher amount of lime additions (Table 5). We find that Fe availability like another heavy metals also decreased with increasing pH of the system and we can say in general pH was most prominent factor that controlled the avilability of Fe. The reduction in Fe availability following liming may be apparently due to abundance of Ca^{++} ions due to lime addition (Datta and Gupta, 1984). The soil reaction, lime content etc. may influence available Fe content by affecting the redox potential and hydration and dehydration of iron compounds in soil and soil like system (Hoshiar Singh, 1964). Another process which can govern the solubility is the adsorption of hydrolysed species and iron itself and predipitaion of metal hydroxides in alkaline media may be proposed as mechanism likely to control the solubilities of iron in soil and soil like system.

Table 5. Periodic changes in Available Fe (ppm) of S-T Mixtures due to liming.

S - T Mixtures	Time (Days)				
	0	15	30	45	60
25A		74.5	82.4	77.8	78.6
25B	90.8	45.4	46	48.6	46
25C		40.2	40.2	42	45.4
33A		78	77.5	80.6	80.4
33B	103.4	55	57	60	57
33C		52.4	50.2	50.8	48
50A		68	65	65	68
50B	132.4	40	41	39	70
50C		27	30	35	33
66A		79.6	80.2	70.4	74.4
66B	153	39	35	40.8	40.6
66C		30.4	36.2	30	25
75A		81.6	75.2	78.4	78.2
75B	180.2	45.6	38	30.4	25
75C		34.8	35	29.6	20.4
100A		50.6	48.8	45.4	43
100B	212.8	45.6	35.4	30.2	24.5
100C		39	31	28	24

The findings of the present study suggest that liming at proper dose is effective in not only rectifying the toxicity of heavy metals but also improving the quality of the degraded land due to copper tailings generated from surface mining and ore processing. However, further detail studies are needed to have a better understanding of the problem. Liming necessarily produced better results but it is to be studied upto the level where plants can grow properly. Although liming reduced the concentrations of various heavy metals significantly but still the metal concentration in tailings amended soil remained in the toxic range. Therefore, a mechanism has to be developed to reduce the concentration of heavy metals further. Moreover few other amendments such as Sewage Sluge, Fly Ash, Pressmud etc. could be tried in different proportions either individually or collectively to assess their suitability for land application. A long term term field experiment based on appropriate statistical design will be needed to evolve any suitable technology for disposal of copper tailings of land applications.

REFERENCES

Bache, B.W. 1973. Reported by Randhawa, N.S. 1974. Micronutrients in soils. Paper presented in the summer institute on secondary and micronutrient elements held a PAU, Lidhiana.

Bandopadhya, A.K. and Adhikari, M. 1975. Technology, 12(1) : 75- 76.

Barber, S.A. 1967. Liming materials and practices in R.W. Pearson and F. Adams (Eds.) Soil Acidity and Liming.

Bradshaw, A.D. 1952. Nature 169 : 1098.

Chatterjeė, R.K. and Dass, S.C. 1964. J. Ind. Soc. Soil Sci. 12 : 297.

Coleman, N.T., E.J. Kamprath and S.B. Weed 1958. Adv, Agron. 41 : 475.

Datta, M. and Gupta, R.K. 1984. J. Ind. Soc. Soil Sci. 32: 450.

Ganjir, B.L., Sinha S.B. and Bhargava B.S. 1973. J. Ind. Soc. Soil. Sci. 21 : 447.

Grewal, J.S., Chaman Lal and Randhawa, N.S. 1969. Ind. J. Agri. Sci. 39(8) : 877.

Gupta, U.C. 1970. Can. J. soil Sci. 114 : 131.

Hoshiar singh, 1964. M.Sc. Thesis, PAU, Hassar.

Ivarson, K.C. 1973. Can. J, Soil Sci, 53 : 315.

Iyengar, S.S., Martenes, D.C. and Miller, W.P. 1981. J. Amer. Soil Sci. Soc. 45 : 735.

Kinniberge, D.C., Jackson, M.L. and Syers J.K. 1976. J. Amer. Soil Sci. Soc. 41 : 796.

Mclaren, R.G. and Grawford,, D.V. 1973. J. Soil Sci. 24(2) : 172.

Pearson, R.W. and Adams, F. (eds.) 1967. Soil Acidity and Liming.

Pulford, J.D., Limber, A.J. and Duncan, H.J. 1983. Int. Conf. Heavy Metal in the Environment, Heidelberg.

Rai, M.M., Mishra, M.K. 1967. J. Ind. Soc. Soil Sci. 15 : 48.

Saeed, M. and Fox R.L. 1977. Soil Sci. 124 : 199.

Shoemaker H.E., Mclean, E.O. and Pratt, P.F. 1961. Proc. Soil Sci. Amer. 25 : 274.

Sinha, M.K., Dhillon, K.S. and Dayanand, S. 1978. Aust. J. Soil Res. 16 : 19.

Singh, M. 1971. Ceoderms 5 : 179.

Smith, P.F. 1968. Proc. Soil Crop Sci. Fla. 28 : 97.

CHAPTER - 28

ROLE OF NON-GOVERNMENT ORGANISATIONS IN MANAGEMENT OF WASTELAND - A CASE STUDY

R.B. Prasad and K. Sushma
S.M. College, Bhagalpur - 812 001

INTRODUCTION

Indian society is well known for its volunatry participation in nation building since time immemorial. Ever since independence most of the schemes and programmes of the five year plans have laid emphasis on involvement of voluntary organisations to realise the goals of planned development. The setting up of Central Social Welfare Board (CSWB) as early as 1953 is a testimony to the fact that the government recognized the role which can be played by voluntary agencies particularly in social development (Sharma, 1997). These agencies have the desired elements of dedication, flexibility and rapport required for bringing social change.

NGOs cover a very wide spectrum of activities which include creating awareness regarding social problems, providing services and enhancing capabilities of communities for keeping the healthy environment. They also play an effective role in advocacy, social mobilization and community empowerment.

NGOs have experimented with innovations and new models of service delivery to meet diverse needs of the communities.

Besides playing an important role in different development projects they also contribute significantly in management of programmes including wasteland and forest development. Conservation, utilisation and human settlements in the vast wasteland of this country have recently been realised seriously. In minimising the problems of wasteland, establishment of National Wasteland Development Board at national level is one of the important steps. The State of Bihar has also initiated same scheme since 1986. The NGOs at national, state and district level are also contributing their energy in the name of "Society for Promotion of Wasteland Development". These societies are engaged with specific objectives to help and accelerate the development of wastelands.

Viewing these facts the present paper is an attempt of compiling the facts regarding NGOs involvement in management of wasteland development in the new district of Banka adjacent to the divisional head quarter of Bhagalpur.

METHODS ADOPTED FOR STUDY

Methods of survey and contacts with the inhabitants of different blocks of Banka district through activities of registered non- government organisations was taken into account for the investigation. Main emphasis especially on bordering land area of plateau region of Santhal Parganas lying in the district was takenup for recording the wasteland improvement. The facts related to baren land, type of cultivation, forest cover, population net work, literacy and number of village areas lacking formal education in centres were collected through NGOs and other secondary source.

Mukti Niketan, Katoria one of the most active non-government organisations engaged in wasteland management was selected for evaluating its activities in some of the villages of Katoria and Chandan block.

SUCCESS ACHIEVED

Blockwise distribution of wasteland, cultivated land area and dmographic composition in the Banka district have been presented in Table-1. Almost all blocks of Banka consist the area of wasteland totalling about 66,271.54 Acre. Maximum area of wasteland (18,082.48 Acre) lies in Blocks of Katoria followed by Chandan (14,133.00 Acre) and Baunsi (12,987.41 Acre). Irrigated as well as non-irrigated crop field growing all types of Bhadai, Agahani, Rabbi and Germa crops covers 3,79,306.89 Acre of cultivated land out of the 7,55,216 Acre of total surface area of the district. Among these, Katoria share the minimum (31,237.68 Acre) of cultivated land in lack of proper water resources. Demographic composition presented in Table 2 also shows that out of 12,47,875 total population there are 6,53,852 male and 5,94,023 female. Bulk of youth population migrates seasonally in other states in search of livelihood and return back in the monsoon season to get their paddy crop.

Table 1. Block wise population and wasteland area in the district Banka

Name of the Block	Population (1991)			Wasteland Area (in Acre)	Total Land (in Acre)	Total Area of Cultivation
	Male	Female	Total			
1. Banka	64600	58030	122630	8497.00	85696	35724.84
2. Katoria	67369	61308	128677	18082.00	158077	31237.68
3. Chandan	51612	48667	100279	14133.00	114664	31896.09
4. Barahat	45683	50700	96383	–	37600	35024.02
5. Baunsi	62958	56843	119801	12987.41	78832	33273.19
6. Dhoraiya	75525	67583	143108	2004.27	58534	31420.02
7. Rajoun	68147	60110	128257	1544.00	52999	45051.49
8. Amarpur	78322	67914	146236	3115.113	49377	47021.08
9. Shambhuganj	72867	62372	135239	1199.27	52823	49248.10
10. Belhar	66769	60496	127265	4709.00	66614	39410.38
Total	653852	594023	1247875	66271.55	755216	379306.89

Information depicted in Table 2 presents improved status of wasteland by active involvement of Mukti Niketan (NGOs) in different villages of Katoria and Chandan blocks. Domsarni, Sathiyari, Baniakura, Gordhasa, Papreba villages lies in the block of Katoria while Manjhali, Dharhara and Bengalgarh belongs to the Chandan block where NGOs has motivated for wasteland improvement. Services rendered voluntarily by the locsl inhabitants in the first phase have improved the wasteland area to fertile field in the range of 20 Acre to 60 Acre in different localities of the villages by creating man made check dam at suitable places in naturally flowing water seepage (throughout the year) in canals of the nearest hilly track. Cheaper system of lift irrigation as well as creation of 3 feet deep underground water distributory canal system (provided with check valve) in the wasteland has helped much in irrigating facilities to the adjoining areas. Local nurseries for timber and fruit plants have also been raised for distribution of seedlings at no profit no loss basis. All the systems are cared and nurtured fully by the local village lift committee without any financial aid or support from the government agencies.

Table 2. Involvement of Mukti Niketan (NGO) in villages for wasteland management

Sl. no.	Village/ Block	Year of Irrigation Started	Population	Cast domination	Wasteland Area (in Acre)	Cultivation Practice Started
1.	Baniakuar (Katoria)	1995	650	Yadav	60	Wheat, Maize, Rice, Potato, Groundnut & tree plantation
2.	Gordhassa (Katoria)	1997	350	·dav	30	Wheat, Rice, Maiz Sugarcane with tree plantation
3.	Papreba (Katoria)	1997	350	Yadav	30	Wheat, Potato, Rice Spices & plantation of tree
4.	Manjhali (Chandan)	1996	792	Yadav Harijan	60	Rice, Wheat, Maize, Pulses & tree plantation
5.	Dharhara (Chandan)	1996	500	Yadav	50	Rice, Potato, Wheat, Sugarcane & tree plantation
6.	Domsarni (Katoria)	1995	968	Yadav	30	Rice, Wheat, Potato Maize & Tree plantation
7.	Sathiari (Katoria)	1995	968	Yadav	30	Rice, Wheat, Potato Maize & tree plantation
8.	Bengalgarh (Chandan)	1997	850	Santhal Harijan	20	Wheat, Rice, Maize, Arhar & tree plantation

Traditional cultivators unkown of getting any crop of wheat, potato, spices etc. in their areas since century back are now getting such crops with great enthusiasm since 1995. Migration of local youths of the areas have stopped considerably under the changed condition of cultivation and plantation.

It is well realized that NGOs which are involved in various development works also play an important role in making the government plan including wasteland development and forestry programme a success. Earlier in the name of welfare services the development works were done, without the participation of the people or without making them to realise the use of project, people were merely considered as passive beneficiaries, thus the actual benificiaries were least bothered for the success of the government owned programme. As a result forest management is still being considered as the responsibility of the government, resulting thereby a massive destruction of the forest mercilessly by the local people itself. The importance of sensitization to the men, therefore, is very necessary.

Involvement of some NGOs in present report clearly reflects the success of many programmes including wasteland development which is based on the proper sensitization, of the respondent through regular meetings and interaction as well as formal and non-formal education. Present findings are also in conformity with the views of Trivedi, (1993). Dissemination of knowledge for raising the nurseries of economic value trees in different village areas including school campus (field) and their large scale distribution are the prime criteria for wasteland improvement.

CONSTRAINS

In the existing scenerio of our country, there is inadequate interaction between government and NGOs. Both seem to be working simultaneously and in a parallel fashion without much of interface of networking. As things exist today the relationship between the government and NGO generally does not go beyound that of giver and taker'. There are apprehensions, mistrust and suspicion on both sides (Sharma, 1995). The time has come when both within the government and the voluntary sector, prejudice and bureaueratic quibbles of status quo should be narrowed down. We need to rist above sectarian of group loyalities to serve the national goals; we need innovation or a vibrant tommorow (Mukherjee, 1992 & Kumar 1995) that can be achieved only through the healthy interaction of government agencies and non-government organisations.

REFERENCES

Kumar K. 1995. Strengthening voluntary action in India, Review Article. Planning commission, yojana, Oct. : 43-44.

Mukherjee, P. 1992. In Proceedings of workshop on creation of Nationwide Network of NGOs for Accelerated Development 29th April, Govt. of India Planning Commission.

Sharma A. 1995. Critical issues in partnership in M.S. Swaminathan Research Foundation Design for Partnership. Proceedings of consultation on NGO Government Partnership.

Sharma A. 1997. Government and voluntary Action for social Development : Meetings for Parternership. ICCW Journal, Jan-June : 59-63.

Trivedi R.N. 1993. Voluntary organisations as facilitators of communication in Rural Development. In text book of Environmental sciences, Anmol Publi. New Delhi : 300-306

Trivedi R.N. 1993. Voluntary Agencies and their Access to Sensitization and Training. In A textbook of Environmental Sciences, Anmol Publi. New Delhi : 395-398.

UNDP 1993. Human Development Report - 93 Oxford University Press, Delhi.

CHAPTER - 29

PROBLEMS AND PROSPECTS FOR RAPIDLY GROWING WASTE LAND:
A CASE STUDY OF SANTHAL PARGANAS IN BIHAR

Kanak K. Pandey

Centre for Regional Study
TM Bhagalpur University, Bhagalpur-812 007

INTRODUCTION

Waste land essentially connote uninhabited portion of land which usually remain uncultivated. It also implies a wild and disolate region which do not fulfill their life sustaining function for one reason or the other. In the present context, wasteland is meant the tract of land or hill made barren as a result of over exploitation of the forest. At a time when the demand for food is ever on the increase because of growing population, the spread of waste land is constantly reducting the valuable land and its productivity in predominantly aggricultural country like India where every acre of arable land is already under cultivation.

Potentially productive land is often lost because of improper land management practices which ultimately result in soil exhaution, erosion, water logging and increase in soil salinity. Nonetheless, more significantly, the ever increasing patches of wastelands are also formed due to the uncontrolled rate of deforestation continuing at an alarming pace in several countries. including India.

"Nature takes 500 to 1000 years to build an inch of top soil. It is estimated that the total amount of soil erosin is of the order of at least 12 billion tons a year in our nation". The soil erosion also induces land slides, premature siltation of reservoirs - rivers, ponds etc, invasion of noxious weeds and severe loss of the valuable genetic diversity of plant and animal species. Recurrent visitation of floods and drought as well as run off water which would otherwise be stored as ground water are causing immense misery to a large number of population.

Accordign to an estimate, the flood prone area in the plains below the Himalayas got doubled from 20 million hectares in 1871 to 40 million in a decade and presently it might have exceeded over 60 million. It is a common experience that when rain occurs, deforested land with inclined plains tend to lose its top layers. On an average it has been estimated that "about 6,000 million tons of top soil each year, which works out to about 8 tons per capita each year" is lost in our country. Which implies that "Each year 6 million ton of soil nutrients are also swept down to the sea, which amounts to 12 per cent more that all the fertilizers used" in India. Soil erosion caused by deforestation afflicts about 150 million hectares, including 16.7 million hectares in river basins.

Not withstanding the fact that the problem of deforestation has assumed such an alarming proportion in the present day, no major reference of the problem is available before the Moghul or British period. While during the Moghul period, the forests were said to have been felled for the military occupation of the area, the British period was marked with an overpowering commercial interest. In the garb of 'rational' and 'scientific' ideology, the Britishers began to encroach upon the pristine world of the tribes and the forests.

It was argued that the tribals recklessly cut down the forests which needed to be prevented. For the purpose, the British India Government in its memorandum of 3rd August, 1865 restricted the rights of the forest dewelling tribes. Numerous rules and regulations in the name of Indian Forest Act followed. Nevertheless, the main thrust of these acts was to give precedence to agricultural needs over the claims of forest preservation. The Indian Forest Act of 1927 gave absolute and unprecedented legal rights to the state. In general its main focus was on the maximum amount of revenue from felling trees for the state. The enactment did not clearly lay down standards for efficient management, conservation, protection and regeneration of forest resources. On the contrary, at the level of attitude also, the administrators considered the forests as an obstruction to the posterity. It was commonly believed that the removal of forests would add to the class of land paying land revenue. As a result "the whole policy of the time was to destroy the forests with this end in view.

The second half of the 19th century witnessed an accelerated onslaught on the forests. In other words, when the expansion of railways was taken up, the great chunks of forests were destroyed to cater to the demands of railway sleepers. As no supervision was exercised over the forest felling operations, a large number of trees been felled were such whose logs could not be utilised.

After independence, there was an awareness in India that organised steps had to taken to repair the damage that occured due to earlier deforestation. The National Forest Policy was accordingly enunciated in 1952 which envisaged that over one third of India's landed area should be under the forest cover. Further, it was also suggested that 60 percent in the hills and 20 per cent in the plains should be brought under the forest cover. However, the satellite imagery for the period of 1972-1975 to 1980-82 collected by the National Remote Sensing Agency of the Department of Space projected a much more deplorable picture. In a nutshell while India's total area under forest cover was 17 per cent during 1972-75, in 1980-82 figure dropped to only 14 per cent As against the Satellite imagery the ground survey conducted by the Department of Forests and Wild-life puts the figure of forest cover to 23 per cent.

The forests in India were always the community preserve. Every one from the rulers to the ruled had a right to get their needs drawn from the nature's preserve. In other words while the rulers utilised the environ for hunting expeditions or even hiding during the time of crises- as born out by the mythological accounts of several classics- the saints and seers wandered in forests in order to seek high points of knowledge and meditations. While on the one hand forest dwellers depended entirely on the forests for all the their needs, the common folks on the other hand were mythologically prescribed to spend their fourth phase of life as Vanprastha in the tranquility of the forest. Incidentally much of the Indian mythological, religious, cultural and social ethos found their origin in the idyllic seclusion provided by the forests. The deforestations generally tend to obliterate, among other things the entire Indian tradition and vitiate the impacts of modernity.

It may, therefore, be argued that the forest dwellers in conjunction with the whole-social system interacted harmoniously with the forests. There was institutionalised social, cultural and religious arrangement and attachment with the forests. As such there was no separate requirements in the form of legal enactments for the protection of the forests. "But once the people were separated from the forests and excluded from them, under the various older legislative acts, these institutions lost meaning and hence decayed and withered away" leading to a wide spread apathy towards the forest environment.

However, in independent India between 1951 and 1980 about 4.3 million hectares of forest land were sacrificed to various non-forestry purpses. According to an estimate "some 17 per cent of Bihar forests have been consumed by the irrigation department for dams and canals, another 8 percent by mines and industries (Srivastav, 1983). The rate of forest falling has amagingly been so rapid, it has been found that 23.11 per cent of forest in the Santhal Parganas Division of Bihar alone was reduced to only 10.13 per cent during a short span of less than two decades. As a result we find that the per capita ownership of forests in Bihar ranks third from below to 0.05 - the other two states been Punjab 0.01 per cent and Haryana 0.01 per cent (Indian Forest State, 1971).

CASE STUDY

Confining to the Santhal Pargana Division of Bihar, it was found that while implementing land settlement during 1914-28 within its old area. 'Daminikoh' was set apart for Kuraon (slash and burn) cultivation, particularly for the Paharia tribes. Excessive exploitation of the area set apart for the above for years together diminished its fertility and the members of the Paharia tribes, therefore, mounted pressure on the neighbouring non - 'Daminikoh' forests. Even a casual observer passing through the districts of Santhal Parganas Division can witness endless patches of denuded forests. According to a survey conducted during 1974-75 to 1988-89 by the state government, about 22,960 hectares of forests were destroyed due to the wasteful practice of Kuraon. Apart from the above, cutting of wood for fuel, fodder, thatching etc. coupled with cattle grazing and incroachment of agricultural activities in the forest area also contributed to the process of deforestation on the large area. The denudation of forests can also be attributed to the illegal transfer of forest land by the ex-jamindar at the time of jamindari abolition. The jamindar owners in large numbers sold the forest trees to the contractors and the tennant cultivators who in turn recklessly removed the forest covers. Besides this the rapidly growing population put unavoidable pressure on the forest land for the numerous requirements. The growing urbanization and modernization in life style also take the toll of the forests.

In an ICSSR sponsored study conducted by the author in the Santhal Parganas Division of Bihar it has been found that 55.45 per cent of the sampled group of respondents suggested that the process of deforestation was almost complete in the region. It was also widely realised that the extensive deforestation had resulted in the scarcity of fruits, flowers, decrease in rainfall, availability of game birds and animals, roots, shoots, tubers and other such minor forest produces that sustained the ramshackle economy of the predominantly tribal forest dwellers. In addition to these, the respondents also realised that soil and hill erosons were taking place in the area as a result of the removal of the tree cover.

Although there is an acute realisation of the fact that the wanton felling of trees is immensely detrimental to the forest dwelling population of the area. Inspite of this, the tribals living in Santhal Paragans find themselves utterly helpless in preventing the onslaught. Due to wide spread poverty the people in the region mostly struggled hard to meet their both ends. As such, it was difficult for them to take any decisive initiative for prevention of the deforestation or even provide protection to the newly planted saplings under the various deforestation programmes launched by the government from time to time (Pandey, 1994).

In order to foster development oriented deforestation, Forest (conservation) Act of 1980 was passed. It laid down that no forest area can be deforested and given for non forestry purpose without the prior approval of the Central Government. The 1983-84 annual report of the Department of Agriculture and Co-operatives has reported that the above act was being observed in commendable manner. In other words, the Government of India had restricted the release of forest land for the purpose of developmental needs and thus reduced the average annual loss of forest area from about 15,0000 hectares prevalent before the enactment to about 3600 hectares annually thereafter.

The available literature on the impact of deforestation have largely been focussed towards the ecology and environment. The studies focussing its attention towards the socio-economic impacts of deforestation are few and far between.

Notwithstanding the above, on basis of common sense

observations it may, however, be argued that extensive socio-economic consequences of deforestation are been faced. For instance, "about 200-300 man days per year are spent on average by a rural family in search of fire wood" (Baidya, 1984). Further, "Some rural families hesitate to send their children to school because they are needed to collect firewood" (Forestry in India, 1984). People are forced to devote more time to fuel wood gathering and transport than to other agricultural activity. In other words " the opportunity cost of fuel wood gathering is far more enromous (Pandey, 1995).

It has also been observed that the use of wood for packaging tea, fruit etc. also lead to deforestation. "Packaging materials are made available at low prices to rich agriculturists with a marketable surplus. While deforestation invariably leads to firewood and fodder scarcity for the poor, the problem is aggravated by pricing policies which ultimately enrich the rich at the expense of the poor, not by normal market laws but by manipulative pricing.

Further, the socio-economic consequences have been particularly evident in the case of tirbal communities who form about 7 per cent of India's population. It has been described earlier, how from the nineteenth century onwards, the forests were either legally or even illigally taken away from them, reducing them to the status of landless agricultural labour. Obviously, the reduction of tribal people to this status was the major cause of their predicament in the contemporary situation and it largely attributed to the deforestation.

While concluding the discussions made in the preceding pages, one may in no uncertain terms observe that the problems of deforestation can be combated only through a massive effort towards afforestation. The policy makers, planners, environmentalists and all other such groups of people are in agreement on this issue. Yet the various programmes of afforestation undertaken so far, particularly in Bihar, have met an avertable doom in the wake of lack of awareness among the people, half hearted political and ecological leadership and innumerable loop holes or corruptions in the implementation agency. The problem of deforestation and the consequent wasteland formation has, however, come to a stage where any form of delay in reversing the process tantamounts to giving an invitation to not only ecological and environmental crisis but also an unreparable socio-economic damage to the mankind as whole.

REFERENCES

Baidya, K.N. 1984. Firewood shortage; eco-climatic disasters in the third world. International Journal of Environmental Studies, Vol.22.

India, Deptt. of Environment, Forestry in India, 1984. An assessment of environmental perspective, New Delhi.

Indian Forest Statistics - Bulletin No.-11. 1971. Central Forestry Commission, Govt. of India.

Pandey, K.K. 1995 Domestic Energy Needs: Patterns of fuel consumption and sources of meeting the requirement in Rural Development in India Ed. U.M. Jha, Anmol Publications, New Delhi.

Pandey, K.K. 1994. Social Forestry - An ICSSR sponsored study Report.

Shrivastava, A. 1983. Present policy making life diffuclt to tribals. The Statemetn.